Advanced Introduction to Digital Society

Elgar Advanced Introductions are stimulating and thoughtful introductions to major fields in the social sciences, business and law, expertly written by the world's leading scholars. Designed to be accessible yet rigorous, they offer concise and lucid surveys of the substantive and policy issues associated with discrete subject areas.

The aims of the series are two-fold: to pinpoint essential principles of a particular field, and to offer insights that stimulate critical thinking. By distilling the vast and often technical corpus of information on the subject into a concise and meaningful form, the books serve as accessible introductions for undergraduate and graduate students coming to the subject for the first time. Importantly, they also develop well-informed, nuanced critiques of the field that will challenge and extend the understanding of advanced students, scholars and policy-makers.

For a full list of titles in the series please see the back of the book. This is also available on https://www.elgaronline.com/ and https://www.advancedintros.com/ for Elgar Advanced Introduction in Law.

Advanced Introduction to

Digital Society

MANUEL CASTELLS

Wallis Annenberg Chair in Communication Technology and Society, University of Southern California, USA

Elgar Advanced Introductions

Edward Elgar
PUBLISHING

Cheltenham, UK • Northampton, MA, USA

Published by
Edward Elgar Publishing Limited
The Lypiatts
15 Lansdown Road
Cheltenham
Glos GL50 2JA
UK

Edward Elgar Publishing, Inc.
William Pratt House
9 Dewey Court
Northampton
Massachusetts 01060
USA

A catalogue record for this book
is available from the British Library

Library of Congress Control Number: 2024934597

Printed on elemental chlorine free (ECF)
recycled paper containing 30% Post-Consumer Waste

ISBN 978 1 80392 111 2 (cased)
ISBN 978 1 80392 113 6 (paperback)
ISBN 978 1 80392 112 9 (eBook)

Printed and bound in the USA

Para Isidora, el amor de mi vida.

Contents

List of figures   viii
Acknowledgements   ix

1   Introduction: a digital society   1

2   We communicate, therefore we are   9

3   End of privacy? The Surveillance State and
    informational capitalism   31

4   The digitalization of financial markets: from
    derivatives to cryptocurrencies   52

5   Teleworking and the networked metropolis   65

6   Human learning, computer learning, AI learning   79

7   Digital divides: territory, gender, age, class,
    ethnicity, cultures   93

8   Networked social movements   110

9   Social media and political polarization   122

10 War and peace in the time of digital machines   132

11 Conclusion: the digital society and the network
    society   145

Index   148

Figures

1.1   Percentage of population that are mobile-cellular
telephone subscriptions and Internet users, by region   2

1.2   Internet of Things (IoT) and non-IoT active device
connections worldwide from 2010 to 2025 (in billions)   3

1.3   Location of servers for main cloud providers
offering infrastructure as a service (IaaS)   4

1.4   Language and image recognition capabilities of AI systems   5

2.1   Time spent per day with digital versus traditional
media in the United States from 2011 to 2023   10

2.2   The media landscape   18

2.3   Estimated circulation of US daily newspapers   21

2.4   Book purchasing: e-books vs printed   22

2.5   The PayPal mafia, key members   23

3.1   Dates when collaboration with NSA began for each
provider   35

4.1   Comparative analysis of global financial values (in
US$ billion)   53

7.1   Scatterplot comparing Internet use and Gini score   101

7.2   Share of websites using selected languages
compared to estimated share of Internet users
speaking those languages   103

Acknowledgements

All books are shared endeavors under the responsibility of the author. This one is no exception. Thus, I want to name and thank the main persons that have contributed to this work in its final form.

My first and foremost gratitude is to Juan Ortiz Freuler, Wallis Annenberg Graduate Fellow, for his truly outstanding research assistance in the elaboration of this book. Furthermore, he has been my interlocutor throughout the long process of analyzing and making sense of the information we have been gathering. Many of his ideas have inspired my writing, albeit I am the only one responsible for the possible shortcomings of the work submitted to your attention. Additional excellent research assistance was provided by my student Marley Randazzo. The students of my graduate courses at the Annenberg School of Communication and Journalism, University of Southern California (USC), particularly in Comm 647 (Network Society) and Comm 670 (Culture and Economy), have also informed with their research papers some of the issues I cover in this book. It goes without saying that their work is fully cited in the references of the relevant chapters.

My efficient administrative assistant Ms. Pauline Martinez at USC has been in charge of the coordination of the whole project and of the preparation of the manuscript with her usual diligence and professionalism.

Ian Tuttle has been, once again, the editor who makes possible that my books, and this one in particular, reach my readers in proper English.

I express my heartfelt recognition to all the contributors that I have cited and to many others, including my colleagues in California, in England, in China, in Mexico, in Argentina, in Portugal, in Costa Rica, and in Spain who have also helped my understanding in our interactions. I am particularly indebted to Jonathan Aronson, Jonathan Taplin, Ernest Wilson, and Geoff Cowan at the USC Annenberg School of Communication and Journalism. My special acknowledgment to my Dean, Willow Bay, for her unwavering support to my work at the School. And I am also intellectually indebted to my colleagues and friends, neuroscientists Antonio Damasio and Hanna Damasio at USC. To John Thompson at the University of Cambridge. To Jerry Feldman at the University of California-Berkeley. To Martin Carnoy at Stanford University. To Imma Tubella, David Megia, Arnau Monterde, and Mireia Fenández-Ardèvol at the Open University of Catalonia. To Cui Baoquo at Tsinghua University. To Fan Dong at Zhejiang University. To Mr. Ren, CEO of Huawei. To Mr. Ma, CEO of Tencent. To my research assistant and former doctoral student at USC, Yuehan Wang. To Carmen Rodriguez Armesta in Mexico. To Fernando Calderon in Buenos Aires. To Gustavo Cardoso in Portugal. To Isidora Chacon in Costa Rica, whose ideas on education transformed my thinking on the matter. And to my colleagues in the Spanish Government from whom I learned the policy implications of the digitalization of society.

As is the case for most of my current research, I have benefitted from the institutional and financial support of the Annenberg Foundation, as well as from the Provost Office at the University of Southern California.

As for the worthiness of all these contributions to the completion of this book, only you, respected reader, will be the judge.

Pacific Palisades, California, November 2023

1 Introduction: a digital society

We live now in an almost entirely digitized society. According to an article published in 2011 in the journal *Science*, whereas in 1986 less than 1% of the world's mediated information was stored in digital format, by 2007 this had reached 94% (Hilbert & López, 2011), and by 2014 it was reaching a staggering 99.5% (Hilbert, 2015). Since producing, storing, and processing information is a key component of all forms of life, it can safely be concluded that digitally producing, storing, and exchanging information shapes the forms of human organization that we call society.

The digital format of information has allowed for an explosion of global communication enabled by Internet protocols and digitized telecommunications. In the world at large, the number of Internet users grew from 2.6 million in 1990 to 5.3 billion in 2022, and mobile subscriptions went from 23,500 in 1980 to over 8 billion in 2020, in a planet of about 8 billion people (World Bank, 2022).

In Figure 1.1 we can observe the growth of mobile phone subscriptions and Internet users in the world by region.

The diffusion of digital communication is the fastest of any technology. In terms of reaching 50 million users: It took 64 years for airlines to do so. It took automobiles 62 years, 50 years for the telephone, 46 years for electricity, 22 years for television, 14 years for computers, 12 years for mobile phones, 7 years for the Internet, 4 years for Facebook (Desjardins, 2018). By 2023, it took just 2 months for ChatGPT to reach 100 million users (Milmo, 2023).

Massive, accelerated digital information creation results largely from the expansion of Internet uses and the growth in the number of Internet users. Thus, in 2021, in one minute on the global Internet, there were 500 hours of content uploaded to YouTube, about 200 million emails sent, 695,000 stories shared over Instagram, 5,000 downloads of the TikTok

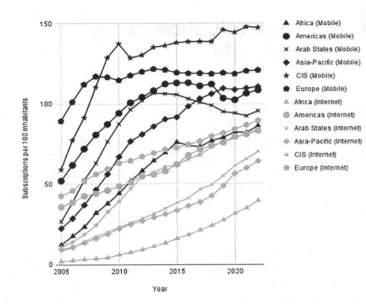

Notes: a). There are more subscriptions than people in many regions because some own multiple connected devices (e.g. work and personal; or tablet and phone), while others own multiple SIM cards to take advantage of different pricing promotions, or to ensure coverage while moving across the territory. b). The region CIS stands for the Commonwealth of Independent States. It is a political and economic organization that was formed in December 1991, after the dissolution of the Soviet Union. The CIS consists of 10 member states, namely Armenia, Azerbaijan, Belarus, Kazakhstan, Kyrgyzstan, Moldova, Russia, Tajikistan, Turkmenistan, and Uzbekistan.
Source: Juan Ortiz Freuler using data from the International Telecommunications Union (UN).

Figure 1.1   Percentage of population that are mobile-cellular telephone subscriptions and Internet users, by region

app, 28,000 subscribers watching Netflix, 2 million swipes on Tinder, and US$1.6 million spent online (Lewis, 2021).

Not only are people generating digital content for other people, but machines are generating digital data for other machines that are connected over the Internet. These are the smart home devices, but also the connected cars and networked industrial equipment. This Internet of

Things (IoT) is growing rapidly, producing a massive amount of digital data that is processed by other machines and which no humans will ever see. These devices are now quickly outpacing human-centered Internet devices, like smartphones, laptops, and computers in terms of active Internet connections, as shown in Figure 1.2.

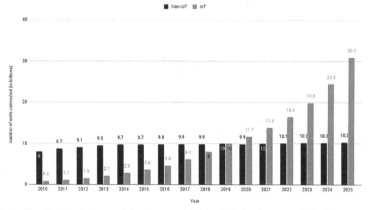

Note: Non-IoT includes mobile phones, tablets, PCs, laptops, and fixed line phones. IoT includes all business-to-business (B2B) and consumer devices connected. Examples of IoT connections include connected cars, smart home devices, and connected industrial equipment. In comparison, non-IoT connections include smartphones, laptops, and computers, with connections of these types of devices set to amount to just over 10 billion units by 2025—three times fewer than IoT device connection.
Source: Author.

Figure 1.2   Internet of Things (IoT) and non-IoT active device connections worldwide from 2010 to 2025 (in billions)

The result is an immense flow of human and computer data that has been accumulating over the years and keeps growing by the minute. This system is only made possible by the rise of cloud computing.[1]

---

[1]   Cloud computing refers to the delivery of computing services over the Internet or "cloud." These services include servers, storage, databases, networking, software, and analytics. It allows users to access these services without needing physical infrastructure, such as servers and data centers, on their own premises. The cloud service provider (e.g., Amazon Web Services,

Expenditures related to the investment in this infrastructure rose from US$61 billion in 2009 to US$178 billion in 2021 (Vailshery, 2022b). While in 2015, 30% of all corporate data were stored in the cloud, the proportion reached over 60% in 2022 (Vailshery, 2022c). The critical role of mega-servers and their geographical concentration raises major issues in terms of countries' sovereignty, as data centers have become major factors of wealth and power. Figure 1.3 shows the unequal geographical distribution of key cloud infrastructure by region in 2018.

Note: Data represents only infrastructure as a service (IaaS) providers and does not include software as a service (SaaS) providers. IaaS is what a company that wants to outsource its data centers would rely on (it does not include data centers managed directly by a company like Microsoft or Facebook to provide services directly to its users). Data only includes the data center information from Alibaba, Amazon Web Services (AWS), Google Cloud, IBM, Interoute, Microsoft Azure, Oracle Cloud, and OVH. Zones are defined as discrete data centers or data center sites (Microsoft's Azure regions are equivalent to AWS and Google's availability zones in this figure).
Source: Author.

Figure 1.3   Location of servers for main cloud providers offering infrastructure as a service (IaaS)

Microsoft Azure, Google Cloud) manages and maintains the infrastructure, allowing clients to focus on using the computing resources for their specific needs. Cloud computing offers scalability, flexibility, cost-effectiveness, and ease of use to its customers, while driving an overall consolidation of the digital markets through abusive gatekeeping of computing resources and horizontal integration by incumbent cloud providers.

The acceleration being described in this text does not concern just the *diffusion* of existing technologies. The *capabilities* of new digital technologies are also accelerating, powered by the growing accumulation of data, novel techniques developed to process such data and learn from it, and available computing power. In Figure 1.4 we can see the quickening pace at which the capabilities of different artificial intelligence (AI) models have achieved human-like performance. Whereas, starting in 1998, it took researchers 18 years to have AI achieve human performance in handwriting recognition, by 2016, it took 4 years for researchers to have AI achieve such human performance in reading comprehension.

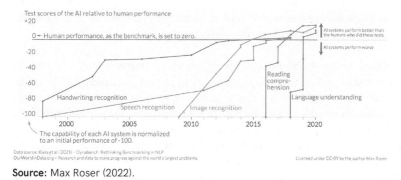

Source: Max Roser (2022).

Figure 1.4   Language and image recognition capabilities of AI systems

Moreover, by 2023, Generative Pre-Trained Transformer (GPT) models were performing better than most humans on most standardized tests (like SAT, LSAT, GRE), with major leaps in progress achieved in just a matter of months (Eloundou et al., 2023). The explosion of Large Language Models (LLMs), most of them managed by private companies, is quickly transforming the landscape of AI applications in education, research, health services, transportation, warfare, management, and decision making, as I will analyze in the following chapters. In a problematic trend, academic research labs are now lagging far behind the private sector, as the percentage of AI PhDs hired by the industry jumped from 25% in 2004 to 73% in 2020 (Ahmed et al., 2023).

Together with the development of AI, the other major recent technological transformation has been the improvement of digital connectivity, with the diffusion of 5G connections, and ongoing progress that will shape what will be known as 6G. The new technologies of connectivity substantially increased the speed of transmission, the volume of data that can be communicated, and a reduction in the latency of communication.[2]

Yet, perhaps the most significant technological change in progress is the coming of age of quantum computing, capable of unprecedented speed and volume of calculation, that is expected to reach the commercial stage around 2030. China defined quantum computing as a key priority in its 2021 Five-Year Plan (Creemers, 2022). The simulation capabilities of quantum computing provide a major tool for strategic decision making in business, in government, and in the military, beyond its promise in the exploration of the new frontiers of science.

These major technological breakthroughs, considered as a cluster of interactive components, are expected to impact decisively on the information processing occupations, as well as services at large. However, this is not necessarily a cause for future mass unemployment. We know from history that rather than phasing out employment, technology replaces humans in the performance of specific tasks, typically reshaping human time allocation toward more skilled activities. However, unless there are protective policies, some professions could be negatively affected. A case in point is the attempt by Hollywood studios to record the images and voices of actresses and actors in order to reproduce their performance by AI-powered virtual characters, thus phasing them out of production without adequate compensation. Similarly, writers could be reduced to becoming editors of scripts written by GPTs trained on the prior work of these same writers. The Hollywood artists and writers' strike in 2023 signaled determined resistance against the substitution of AI for humans in a wide range of activities. The ultimate outcome will not depend on technology but on power relationships.

---

[2]   Latency refers to the delay between an action and its response in telecommunications. In 5G networks, achieving low latency is crucial to provide faster response times and better performance for applications such as remote medical care, virtual reality, and online gaming. It enables near real-time interactions. The 5G standard sets a target latency of less than 1 millisecond, which is significantly lower than the average latency of 4G networks. To achieve low latency in 5G, improvements in network architecture and advanced technologies such as network slicing and millimeter wave frequencies are required.

Provided that the training and education system is transformed accordingly, and that governments and companies assist workers in the transition while respecting their rights, the impact could be positive in terms of productivity and quality of life. Fears of negative disruption by the acceleration of digitalization may be appeased on the basis of social science-based knowledge of specific effects, sector by sector. Institutions will have to design policies to cope with the new challenges of what could be a quantum leap of human creativity, or else blindly jump into an uncertain technological galaxy. This is what this book attempts to explore, free of ideological prejudice.

References

Ahmed, N., Wahed, M., & Thompson, N.C. (2023) "The growing influence of industry in AI research," *Science*, 379(6635), pp. 884–886. https://doi.org/10.1126/science.ade2420.

Creemers, R. (2022) "Translation: 14th Five-Year Plan for national informatization – Dec. 2021," *DigiChina*. Available at: https://digichina.stanford.edu/work/translation-14th-five-year-plan-for-national-informatization-dec-2021/ (accessed on 20 April 2023).

Desjardins, J. (2018) "How long does it take to hit 50 million users?," *Visual Capitalist*. Available at: https://www.visualcapitalist.com/how-long-does-it-take-to-hit-50-million-users/ (accessed on 21 March 2023).

Eloundou, T., Manning, S., Mishkin, P., & Rock, D. (2023) "GPTs are GPTs: An early look at the labor market impact potential of large language models," arXiv. Available at: http://arxiv.org/abs/2303.10130 (accessed on 20 March 2023).

Hilbert, M. (2015) "Quantifying the data deluge and the data drought," *SSRN Electronic Journal*. https://doi.org/10.2139/ssrn.2984851.

Hilbert, M., & López, P. (2011) "The world's technological capacity to store, communicate, and compute information," *Science*, 332(6025), pp. 60–65. https://doi.org/10.1126/science.1200970.

Lewis, L. (2021) "A minute on the Internet in 2021," *Statista Infographics*. Available at: https://www.statista.com/chart/25443/estimated-amount-of-data-created-on-the-internet-in-one-minute (accessed on 20 April 2023).

Milmo, D. (2023) "ChatGPT reaches 100 million users two months after launch," *The Guardian*, 2 February. Available at: https://www.theguardian.com/technology/2023/feb/02/chatgpt-100-million-users-open-ai-fastest-growing-app (accessed on 20 April 2023).

Roser, M. (2022) "The brief history of artificial intelligence: The world has changed fast – what might be next?," *Our World in Data*. Available at: https://ourworldindata.org/brief-history-of-ai (accessed on 20 April 2023).

Vailshery, L.S. (2022a) "Global IoT and non-IoT connections 2010-2025," *Statista*. Available at: https://www.statista.com/statistics/1101442/iot-number -of-connected-devices-worldwide/ (accessed on 20 April 2023).

Vailshery, L.S. (2022b) "Percent of corporate data stored in the cloud 2022," *Statista*. Available at: https://www.statista.com/statistics/1062879/worldwide -cloud-storage-of-corporate-data/ (accessed on 20 April 2023).

Vailshery, L.S. (2022c) "Global cloud and data center spending 2021," *Statista*. Available at: https://www.statista.com/statistics/1114926/enterprise-spending -cloud-and-data-centers/ (accessed on 20 April 2023).

World Bank (2022) *World Development Indicators|DataBank*. Available at: https://databank.worldbank.org/indicator/NY.GDP.MKTP.KD.ZG/1ff4a498/ Popular-Indicators (accessed on 8 May 2021).

2 We communicate, therefore we are

Humans are social animals that develop their consciousness and organize their lives by communicating with each other. Our neural networks connect with the neural networks of other human individuals and with the networks of our natural and cultural environment. Communication is the construction of meaning through the exchange of information. Thus, a technological revolution focused on information and communication necessarily induces powerful effects in the entire human experience, in close interaction with the cultures and institutions of societies in all their diversity.

The twentieth century saw the rise of mass communication, which was characterized by one-directional messages sent from a limited number of senders to a very large number of receivers (McChesney, 2007; Neuman, 2016). With the advent of the Internet in 1969, and the gradual deployment of interactive digital communication networks, a new form of communication emerged: mass self-communication (Castells, 2009). It is "mass" because it has the capacity to reach a global audience, surpassing in its reach any prior form of communication. It is "self" because the messages can be sent, received, selected, retrieved, combined, and interacted with by both sender(s) and receiver(s). Furthermore, relentless multiplicity of messages converges in a hypertext that becomes the evolving frame of reference for all communicative practices. However, the autonomy of the self-communicated being is relative. It is mediated by social organizations, and by the owners and controllers of the digital networks of communication that increasingly manage messages on the basis of algorithms whose parameters are unknown to the communicating subjects. As the industrial technology and the industrial society shaped mass communication, originating the mass media, communication technology and the network society patterned mass self-communication. The diffusion of digital networking technology has been the fastest of any communication technology in human history, as shown in the introduction to this book.

Mass media is being overtaken by what has been labeled "social media"[1] in terms of time spent, as shown in Figure 2.1.

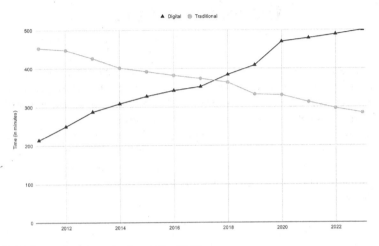

Note: Data refers to projections after 2020.
Source: Designed by Juan Ortiz Freuler based on data aggregated by Statista, 2023b.

Figure 2.1   Time spent per day with digital versus traditional media in the United States from 2011 to 2023

The new communication sphere transcends the boundaries of space and time. It is local and global, multimodal, synchronous and asynchronous at the same time. In terms of human-to-human interaction, the key technology was the development of the so-called "smartphone" by Steve Jobs and his team at Apple in 2007, that is, a handheld computer with an Internet-focused communicating device. It distributes messages and data over communication networks from anywhere to anywhere. What is called "mobile communication" does not necessarily operate on the

---

[1]   Social media is an ambiguous term because all media are social. I follow boyd and Ellison in naming these communication networks as "social network sites" (boyd & Ellison, 2007). In fact, the proper concept is "digital social network sites," because it is essential to specify the technology of communication.

move: most calls are from workplaces, homes, and schools. The critical feature is that it ensures permanent connectivity (Katz & Aakhus, 2002; Ling, 2004; Castells et al., 2006). With 8.6 billion wireless subscriptions in 2023 (Statista, 2023a) in a planet of 8 billion people, and with 86% of the world population using smartphones, we have crossed a threshold of a social organization that could be conceptualized as the "communication society," as Alain Touraine proposed (Touraine, 2021) rather than the "information society." Internet protocols made globally distributed digital communication possible, much like the electrical engine decentralized energy power in industrial societies. Granted, inequalities in the use of the Internet persist, as I will analyze in Chapter 8. Yet, these are inequalities within a shared social structure (the network society) and a shared technological paradigm (the digital society).

Sociability 3.0

Being social means relating to other people at different levels of intimacy and frequency. Sociologists usually differentiate between weak ties (such as occasional acquaintances) and strong ties (family, long-standing friendships, love relationships) (Wellman et al., 2001). Sociability 1.0, in the pre-industrial age, was largely determined by family, mainly by extended family, and contiguity at work or residence. With large-scale urbanization prompted by industrialization, intimate relationships were limited to a close-knit group, while social interaction diffused over space and time, with the separation between work and residence. Participation in society at large took the form of voluntary associations and segmented interest groups. It was a transition from "community" to "association" (Fischer, 1976; Tönnies, 2001 [1887]). This form of sociability 2.0 included the previous form, but became more comprehensive. The formation of opinions and values that would guide behavior was increasingly dependent on mass media, from the print press to radio and television. Telephone interaction connected individuals (Fischer, 1994).

With the Internet and the explosion of social media, a new form of sociability, which we could call 3.0, gradually took shape (Castells, 2001; Katz & Rice, 2002; Castells et al., 2003; Baym, 2015). Strong ties were usually created by face-to-face interaction or through family and other forms of constructed social life. But Internet-based communication emphasized strong ties while making possible, with little effort, the exploration and maintenance of weak ties (Rainie & Wellman, 2012). The coming of this new sociability was clouded by nostalgia for the loss of "authentic sociability," denying Internet interactions any meaningful role in the expression of deep emotions. Moreover, cultural resistance to change blamed online sociability for destroying the previous social life. With the Internet, as has usually been asserted in the media based on anecdotal evidence, people were considered to be isolated, lonely, depressed, and ultimately alienated (Wolton, 2000). We entered full speed into a new society failing to recognize what it was, and attributing to the Internet responsibility for our grave social ills, as had also been the case earlier with the diffusion of television (Postman, 1986), albeit Umberto Eco challenged this assumption (Eco, 1984). Fears focused on the potentially damaging consequences for children and youth. They were the Internet generation, sentenced to be mentally impaired, incapable of learning, and ultimately prone to becoming dangerous nerds because they were living in a communication environment that their parents did not understand.

Fortunately, we have good social science and dozens of research institutes for the social study of the Internet around the world, such as the Oxford Internet Institute, the Harvard Berkman Klein Center, the Pew Research Center in Washington or the Internet Interdisciplinary Institute in Barcelona. A number of studies have shown that, while many youths do have personal problems and feel isolated, the use of the Internet was alleviating rather than causing these problems. It has been shown that off-line sociability and online sociability were cumulative, not exclusive from each other. That is, the more you are social face-to-face, the more you are social on the Internet, and the more you socialize online, the more you broaden your overall sociability (Castells et al., 2003; Hampton, 2004; Rainie & Wellman, 2012). In synthesis, the results of academic research show that a new form of intense sociability emerges.

When people have feelings of isolation and alienation, being social on the Internet often helps them to cope with these feelings. Rather than unhappiness resulting from the use of the Internet, the important study by Nahoi Koo, among others, showed the contrary. Using data from the University of Michigan World Values Survey on a global representative sample, she observed that the intensity of Internet use correlates positively with the indicators of life satisfaction (Koo, 2017). There is an explanation for this finding that coincides with other studies on the

Internet: Internet increases overall sociability. And the use of the Internet empowers people who are disempowered, particularly women, ethnic minorities, and broadly geographically dispersed like-minded people (BCS, 2010). In her structural equation model to explain subjective satisfaction of Internet users, Koo introduced the two key variables that increase personal well-being: density of social relationships and empowerment as a result of access to information. Both factors are stimulated by the use of the Internet (Graham & Dutton, 2014). Furthermore, against the image that was created by some early studies about the predominance of fake identities on the Internet, grouping in social networking sites, as in Facebook, is based on affinity and convergence of interests.[2]

In fact, the available evidence shows (Persily & Tucker, 2020) that politically motivated groups are built on some common assumptions of ideological values and political positions, a phenomenon at the origin of political polarization in the networks and in society. It could well be that some users hide their real names, but not their ideas and identities, since the purpose of participating in the interaction is precisely to find companionship in their orientations (Baym, 2015). Faking names is not equivalent to faking identities. Exposure of who they are is essential for most users. In part, it is a narcissistic practice that is pervasive in digital media. But more importantly, it is indicative of a search for soulmates that soothes the misunderstanding, isolation, and hostility suffered in social life (Jones, 1997; Yates et al., 2020).

As the use of the Internet expanded to all spheres of social life, it contributed to amplify or even transform some of the most important dimensions of human relationships, such as heterosexual dating and mating. A number of statistical studies in the United States (US), summarized and discussed by Marley Randazzo (2022), found that by 2017, before the Covid-19 pandemic, about 40% of new couples had met online, followed by introduction by friends (20%), meeting at work (10%), school (10%), family (7%), while the number of those meeting in bars was negligible. Of course, the 2020 pandemic accelerated, but did not cause, the trend toward

---

[2]   For instance, if cats formed a group, it would be of little use for a dog to join it, even if it can take advantage of the fact that on the Internet no one knows that you are a dog, as suggested in the now famous *New Yorker* cartoon. In any case, nowadays the networking site knows everything about who is in the group, as I will argue in Chapter 3.

digital dating as a match-making mechanism, creating a large market for digital dating platforms. Contributing factors appear to be the need to evaluate features of potential partners before engaging in a relationship, while retaining the possibility of maintaining simultaneous relationships before making a choice. A number of factors seem to be operating in this new sociability pattern. First, marriage has been delayed (in the US, the median age for men to marry is 30 years and for women it is 28). Second, contrary to public perception, the share of adults (particularly male) aged 18–30 who had not had sex in the last year grew from 10% in 2008 to 23% in 2018 (it should be remembered that this concerns heterosexual sex) (Ingraham, 2019). This could be a result, as we hypothesized some time ago (Castells & Subirats, 2007), of men's reaction to the growing autonomy and consciousness of women. Men may be increasingly cautious before committing, a trend that could explain simultaneously late marriages, less male sexual activity, and greater use of digital platforms of intermediation.[3]

What is clear is that intervening in the midst of a deep cultural transformation of heterosexual romantic practices, digital dating platforms contribute to modifying one of the most important traditional practices of human sociability.

This trend is consistent with another major socio-technical evolution: the Internet has been shown to be a powerful contributing factor favoring the autonomy of individuals vis-à-vis the norms and values that are dominant in institutions (Rainie & Wellman, 2012; Graham & Dutton, 2014). There is no crisis of sociability, but rather the emergence of a new form of sociability, conceptualized by Wellman as "networked individualism." Individuals define their own interests and life projects, but they network with other individuals.

In 2003, I conducted a study based on a large survey of 3,000 people, a representative sample of the population of Catalonia (Castells et al.,

---

[3]   There are cultural differences in the frequency of sexual activity. For instance, in Spain in 2021, the average frequency of sexual activity for men was once a week. In Spain, digital platforms have also increased their role in dating and mating in the last decade. Another element to be considered is that pornography with advanced digital technologies is rising sharply in most countries.

2003). We defined five types of social autonomy: personal, professional, socio-political, cultural, and body-centered. We showed that the five dimensions were statistically independent among themselves. But every one of them correlated positively with the level of autonomy shown by the surveyed individuals, as measured on a scale of autonomous behavior. This is a significant finding because, as Giddens has theorized, the process of individuation is an essential feature of our contemporary societies (Giddens, 1992). Understanding by individuation the capacity of people to define autonomous projects, distancing oneself from the dominant norms of society or organizations. Technologies are adopted and become powerful tools for humans when they fit with the values of a significant number of humans in a given historical context. So, the Internet is a tool of communicative autonomy that fits particularly well in a culture in which values of individuation result in the definition and implementation of self-defined projects by the most autonomous people in society. The Internet was created by a culture of autonomy (Castells, 2001), and, in return, decisively contributes to the strengthening of this culture for autonomous social actors, who are a decisive minority in the exploration of the new frontiers of innovation and social change. Moreover, on the Internet, the users are producers of innovation that shapes social interaction both there and in society. Yet, notwithstanding this relative autonomy of users/producers, Internet platforms are owned and managed by companies that often condition users' interaction through their design features and the algorithms that manage moderation and curation of the circulating content, among other examples.

Digital social network sites

Social interaction on the Internet, for all kinds of purposes besides sociability, is based on digital social network sites. Following the definition by boyd and Ellison (2007: 211), social network sites are "web-based services that allow individuals to (1) construct a public or semi-public profile within a bounded system, (2) articulate a list of other users with whom they share a connection, and (3) view and traverse their list of connections and those made by others within the system." The users of social network sites (popularly labeled "social media") account for the majority of Internet users nowadays. The first large-scale site was Friendster, launched in San Francisco in 2002. Facebook started at Harvard in 2004, then moved to Silicon Valley to benefit from the ecosystem of innovation and its labor pool. In January 2023, worldwide there were thousands of social network sites for all kinds of purposes with 4.7 billion users, or 59.4% of the world's population, thus representing the most important form of extended sociability globally, always coexisting with face-to-face interaction. While 25.9% of world Internet users communicate using the English language, almost three-quarters of users do not use English, including 19.8% who use Chinese and 8% Spanish (Statista, 2022b). Moreover, there is a gradual and relative decline in the dominance of US platforms like Facebook and other Meta networks, due to the increasing weight of Chinese networking sites (such as WeChat, TikTok-Douyin, Kuaishou, etc.), and alternative networks such as Telegram and Signal. Table 2.1 ranks the main social site networks by numbers of users.

The relative ranking of social networks is not as stable as it may appear. It is highly sensitive to changes in management (e.g., initial exodus of Twitter users after it was taken over by Elon Musk), to technological innovation from competitors (TikTok outcompeting Facebook), or to the obsolescence of business models based almost exclusively on targeted advertising (as appears to be the case with Facebook transforming itself into Meta and seeking revenue from the sale of virtual reality headsets, smart sunglasses, and marketplace fees). Ultimately, the transformation of the digital communication industry is impacting substantially the practices in social networks as new technologies, new business models, and new organizational structures emerge.

The convergence of communication and Internet industries: multimedia business networks

Digital communication and the media industry are organized around a handful of very large companies at the core of multimodal business networks, as Arsenault and Castells showed in their study in 2008 (Arsenault & Castells, 2008). However, while the global, networked structure of the industry persists, the nodes of its configuration have changed over time, and continue to change as business diversifies, markets expand, and mergers and acquisition follow the transformation of capital and technology. Technology companies, social media companies, entertainment companies, e-commerce, financial services, and distribution companies

Table 2.1   Most popular social networks worldwide as of January 2023, ranked by number of monthly active users (in millions)

Platform name	Number of monthly active users (millions)
Facebook	2,958
YouTube	2,514
WhatsApp	2,000[a]
Instagram	2,000
WeChat	1,309
TikTok	1,051
Facebook Messenger	931
Douyin	715[b]
Telegram	700
Snapchat	635
Kuaishou	626
Sina Weibo	584
QQ	574
Twitter	556
Pinterest	445

Notes: a). Platforms have not published updated user figures in the past 12 months, so figures may be out of date and less reliable. b). Number of daily active users, so monthly active user number is likely higher.
Source: Author.

mix and recombine constantly trying to reinvent themselves in an industry marked by relentless technological and organizational change.

The data and networks presented in Figure 2.2 should be interpreted with care. First, in sharp contrast with the data from 20 years earlier, Internet companies (Facebook, Amazon, and Google) seem to dominate the multimedia business structure. This is because they have become key nodes for the distribution of third-party content via the Internet, a relevant source of revenue nowadays, and one that offers gatekeeping powers, often abused to gain a competitive advantage or extract fees. An

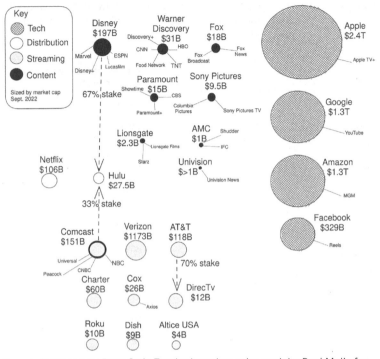

Source: Designed by Juan Ortiz Freuler based on prior work by Rani Molla for Recode (2022) using data from company reports, Lechtman Research group and Recode reporting.

Figure 2.2   The media landscape

example of this is the way in which iOS (Apple) and Android (Alphabet) manage the app store, and the "tax" they impose on sales that take place within their mobile environments, which range from 15 to 30%. Content creator platforms like Facebook (30% of monthly fan subscriptions) and YouTube (45% of revenue from ads on videos), as well as the more professional space of Amazon (20–30% on subscriptions), also extract value from any transaction taking place through their platform (Campbell & Alexander, 2021).

Secondly, using their abundance of capital, these companies are able to invest and network with multiple companies to disintermediate the connection with their users, and to control the production of content. For instance, Amazon or Meta have the financial resources to venture into new projects that may fail or yield results only in the long term. This may be the case for Zuckerberg's Metaverse that, as of 2023, did not appear to be any more successful than the first virtual reality experiment, Second Life, which launched in 2003 and which, by 2010, was going through a series of massive layoffs, resulting in a cut of 30% of its workforce.

Thirdly, market capitalization might be a misleading measure by which to understand this market after media convergence, combined with mechanisms of horizontal and vertical integration, make the listed market capitalization the result of a much broader strategy than simply media. For example, Apple's revenue (and arguably its market capitalization) is preeminently a result of its capabilities as a manufacturer of the iPhone and other devices. Apple, and other tech companies included here, do not publish disaggregated data that might help us better understand or estimate the specific role played by its media industry branches, such as AppleTV streaming entertainment, in their overall market capitalization.

Fourthly, cross-collaboration is still key. Observing the more traditional multimedia segment of the industry, the dominant companies in terms of market capitalization are different: Disney, Comcast, ATT, Sony. Each one of them is part of a conglomerate of other media and social media companies, related in networks of collaboration. For instance, Disney, outcompeted by Netflix for some time, reached parity in terms of subscribers by networking with Hulu, and by diversifying its content lines with different television channels and film studios, besides investing in theme parks and merchandising. Hulu develops content with Disney and networks with Comcast to increase its reach.

Fifthly, a key tech company like Microsoft is not in the graph, in spite of being the second largest company in market capitalization in the world. This is because the failure of its Internet Explorer limited its gatekeeping role in the Internet and in the media industry at large. Yet, its new search engine, Bing, is starting to compete with Google, and it is becoming a giant in the media-adjacent gaming sector. Indeed, Microsoft remains a major source of innovation in the software computing industry and, as such, is a key technological resource for media and communication companies. Its role as a key node will certainly increase as a result of its strategy of investing in AI, mainly through its participation in OpenAI, an organization to which Elon Musk had originally provided funds. The

in the demiurgic power of technology, beyond their diverse political preferences. Although the Silicon Valley entrepreneurs were always driven by their passion for technological innovation, the new generation feels more empowered than ever as shown by their risk-taking strategies aimed at "disrupting" a wide variety of sectors: virtual pseudosocial media (Zuckerberg), automated transportation and space exploration (Musk), finance (Andreesen, Thiel) or government, including security and spyware (Thiel).

Their libertarian belief, often contradictory in its application, becomes embodied in their financial and technological capacity to impose their will as individuals, calling on their believers to sever ties with institutions and to assume the power they deserve as superior individuals. Peter Thiel clearly articulates this ideology of authoritarian technocracy on behalf of the individual: "Your mind is software. Program it. Your body is a shell. Change it. Death is a disease. Cure it. Extinction is approaching. Fight it" (Stein, 2014). As for Elon Musk, "[t]he long-term ultimate objective—the holy grail—is: we would like to make life multiplanetary" (Stein, 2014). These are expressions of ideas but they are not without consequences: ideology materializes in technologies by which we live our lives.

In its search for new sources of profit and power, the communication industry moved from capturing audiences to shepherding weak friendships. Then from interacting with other humans to cohabit a multiverse disembodied as avatars, or as a multiplanetary species, thus attempting to construct a new form of sociability: Sociability ∞.

A disinformed information society: fake news

The large majority of people in the US and most digitally advanced European countries obtain their news, and information at large, from digital social network sites and search engines. Meanwhile, readership and revenue for newspapers is declining precipitously. Television viewers are vastly outnumbered by social media users. As a result, the filtering role exercised by professional journalism is fading away. We must differentiate rumors from fake news ourselves. Rumors refer to unverified accounts of some event. Rumors have always existed in communication environments, and history is full of examples of the powerful effects of rumors that distort the actual experience. Fake news, according to some researchers, includes messages diffused by different means that convey erroneous information, willingly or unintentionally. In other definitions, fake news relates to the deliberate spreading of news that is knowingly false or inexact, which is also referred to as disinformation. What is labeled as "deep fake" uses digital technologies, particularly AI, to achieve highly realistic effects in the fabrication of content, such as depicting a violent act or a statement by a prominent politician who appears in person saying what the creators of news find convenient to their cause. For example, in the Brazilian presidential election of 2018, a deep fake portrayed candidate Lula as threatening his opponents.

In the US, in the 2016 presidential election won by Donald Trump, fake news, of different kinds, became highly influential in the dissemination of political information. Trump followers, and to a lesser degree Russian trolls, distributed a massive amount of fake news to damage Hillary Clinton's image, contributing to her defeat. While all communication environments have seen disinformation distributed on a mass scale, Internet networks and digitization have decisively contributed to pollute the public information environment, particularly amplifying fake news and increasing its credibility. This is, first of all, because of the rapid diffusion of Internet messages: virality, defined as "a social information process in which content is shared with often distant networks resulting in a sharp acceleration of social information" (Nahon & Hemsley, 2013). Since most people get their news from social media and online news outlets, forwarding the message is easy, fast, and could be targeted on a massive scale.

Secondly, social bots, with their networks of reach programmed to achieve complementarity and impact during propaganda campaigns, automate and multiply the forwarded messages.

Thirdly, the development of powerful AI algorithms, such as ChatGPT 4.0 and others, sharply increases the capacity to design and distribute credible fake news, blurring the digital information sphere where fact and fiction can hardly be differentiated.

Fourthly, because of the well-known fact that negative messages stick more effectively with viewers, the negativity of fake news constructs a receptive environment for such news. Moreover, over the past decade, more news headlines have become increasingly negative as an acknowledgment that this fuels distribution (Robertson et al., 2023). However, we must also consider that individuals lend greater credibility to messages that fit with their predispositions (Castells, 2009). In other words, targeted disinformation campaigns focus on reinforcing negative perceptions of those that already feel that way. But they also induce doubts in people or voters uncertain in their opinions. Since a key characteristic of contemporary societies is the growing distrust in institutions, including the mainstream media, it follows that communication environments are more receptive to messages that challenge stories published by professional journalism. The declining legitimacy of societal institutions induces a self-fulfilling prophecy: the more there is suspicion about information conveyed by traditional media, the more people are receptive to fake news that eventually confirms their distrust.

However, there are important cultural differences in the acceptance of negative fake news depending on the country and on the level of education. The more educated a person is, the more likely that she will develop a capacity to critically engage with content, and vice versa. Also, national differences are conducive to a greater acceptance or not of the "official story," depending on the degree of consensus between citizens and their institutions. This is not confined to political or ideological opinions, but includes all cognitive dimensions of social life. For instance, during the Covid-19 pandemic, a significant segment of many societies as educated as, for instance, France, lent credibility to rumors, as well as to fake news, that challenged the belief in scientific consensus and led to distrust in vaccines. This disinformation was very costly in human lives, and conditioned the feasibility of many health policies. In China, 40% of people over 65 refused to be vaccinated to the point that the government resorted to massive, lengthy confinement to control the spread of the virus, ultimately provoking resistance that aggravated the pandemic. Conspiracy theories, such as attributing to the vaccines a capability of implanting chips in our body, populated social networks, caused social anxiety, and ultimately resulted in thousands of deaths. The crisis of institutional trust was amplified in the digital social networks. In so doing, the social "truth" disappeared in the minds of millions of people, undermining the capacity of humans to live together under common social norms. For instance, the calling into question of science, which has been a fundamental factor underlying the historical experience of social progress and shared well-being, is largely a result of disinformation in the digital social networks. The growing role of fake news in shaping the minds of humans by diffusing such "news" at an accelerated pace is a destructive trend that threatens our survival, both as individuals and as a species.

In this context there are increasing calls for regulation of the digital media environment from many quarters, including politicians, scientists, and some leaders of technology companies. However, these well-intentioned proposals contrast with the reality of the vested interests of capitalism and the State in our time, namely, informational capitalism and the Surveillance State.

References

Arsenault, A.H., & Castells, M. (2008) "The structure and dynamics of global multi-media business networks," *International Journal of Communication*, 2, pp. 707–748.

Baym, N.K. (2015) *Personal Connections in the Digital Age*, 2nd edn. Cambridge: Polity.

BCS (2010) *The Information Dividend: Can IT Make You "Happier"?* Swindon, UK: Trajectory Partnership. Available at: https://www.trajectorypartnership.com/wp-content/uploads/2013/09/BCS_Information_Dividend_Global.pdf (accessed on 18 October 2023).

boyd, d.m., & Ellison, N.B. (2007) "Social network sites: Definition, history, and scholarship," *Journal of Computer-Mediated Communication*, 13(1), pp. 210–230.

Campbell, I.C., & Alexander, J. (2021) "A guide to platform fees," *The Verge*, 22 September. Available at: https://www.theverge.com/21445923/platform-fees-apps-games-business-marketplace-apple-google (accessed on 18 October 2023).

Castells, M. (2001) *The Internet Galaxy: Reflections on the Internet, Business, and Society*. Oxford: Oxford University Press.

Castells, M. (2009) *Communication Power*. Oxford: Oxford University Press.

Castells, M., & Subirats, M. (2007) *Mujeres y hombres : ¿un amor imposible?* Madrid: Alianza Editorial.

Castells, M., Tubella, I., Sancho, T., Diaz de Isla, M.I., & Wellman, B. (2003) *La societat xarxa a Catalunya*. Barcelona: Mondadori, Rosa dels Vents y Editorial UOC.

Castells, M., Fernández-Ardèvol, M., Qiu, J.L., & Sey, A. (2006) *Mobile Communication and Society: A Global Perspective*. Cambridge, MA: MIT Press.

Eco, U. (1984) "Does the audience have a bad effect on television," in U. Eco & R. Lumley (eds), *Apocalypse Postponed*. Bloomington, IN: Indiana University Press.

Fischer, C.S. (1976) *The Urban Experience*. New York: Harcourt Brace Jovanovich.

Yates, S.J., Blejmar, J., Wessels, B., & Taylor, C. (2020) "ESRC Review: Communities and Identities," in S.J. Yates and R.E. Rice (eds), *The Oxford Handbook of Digital Technology and Society*. Oxford: Oxford University Press, pp. 404–425.

3 End of privacy? The Surveillance State and informational capitalism

Throughout history, a fundamental instrument to assert power is the control of information and the monopoly of communication (Castells, 2009). A key mechanism to fulfill these goals is the collection of information on people in an asymmetrical manner through surveillance, where the surveilled are kept in the dark. In informational capitalism (Castells, 2000; 2012), the accumulation of information on human activities and their subsequent commodification is a form of capital accumulation. The appropriation of communication messages is ensured by corporate control over the networks of communication, and government control over the corporations, thus combining their technical and legal capacity to accumulate and distribute information selectively to targeted audiences. A digital society is characterized by a digital hypertext that is at the same time constantly produced and modified and constantly accessed, recombined, remixed, and redirected by the communicating actors. Under the conditions of digital information and communication, there is simultaneously an increasing centralization of information and a decreasing monopoly over communication. There is a decreasing monopoly over communication because of the rise of mass self-communication, as defined in the previous chapter. In this paradigm there are billions of interactive, communicative subjects. However, information is increasingly centralized in the servers of a handful of Internet-based companies, and in government servers. In fact, a key factor that enables the centralized control of information is precisely the decentralized, multiple networks of communication where everybody and everything exchange messages and information. The technological ability to surveille, track, store, and analyze these multiple processes of communication, and the accumulated hypertext resulting over time from the exchange of information, represents a substantial leap in the control over information. The technological and organizational capacity required to collect data at a planetary scale requires considerable resources, particularly political power and corporate wealth, which are in the hands of State agencies and a few private corporations.

Meanwhile, the possibility of communicative autonomy increases the capacity of people and organizations to exchange information over Internet networks. Certainly, these networks are surveilled, mediated by companies' algorithms and used by corporations and by the State to gather information. But in general terms, the flow of information is not interrupted precisely because unfettered communication provides the opportunity to access and collect information by the institutions and companies able to do so.

Surveillance is certainly a form of invasion of privacy. But the most significant threat to privacy, defined as the right of people not to disclose information about their lives without their explicit consent, does not come from governments but from corporations, and precisely from the Internet and communication corporations (Castells, 2001). In this case the breach of privacy derives from a business model partly based on retrieving and exploiting commercially the information provided by or retrieved from users. Contemporary capitalism is informational capitalism (Cohen, 2019).

Thus, the threats to democracy resulting from surveillance and the threats to privacy resulting from the commodification of our lives signal the persistence in the Information Age of the domination of social life by the two major institutions that structure society: the State and capital.

The rise of a global Big Brother

State surveillance is common to all States. However, digitization has now provided the technological platform for the formation of a globally networked surveillance bureaucracy. At its core there is the US National Security Agency (NSA), which is connected by international agreements, formal or informal, to other similar agencies. The Five Eyes cooperation program between the US, the UK, Canada, Australia, and New Zealand exemplified the networking strategy that, over time, strengthened the joint surveillance capabilities of Western governments. The Five Eyes program (1941) expanded to Nine Eyes (late 1940s), incorporating France,

Denmark, Netherlands, and Norway. Later, the agencies of Germany, Belgium, Italy, Spain, and Sweden were added, forming the Fourteen Eyes program (mid 2000s). Ultimately, following the geopolitical configuration of States, the surveillance networks include all NATO countries and their allies, such as Israel's Mossad (Greenwald, 2014). Russia and China developed their own networks centered on their very large surveillance agencies (Russia's FSB and China's MSS) (Gill & Phythian, 2018). Other countries, such as India, Pakistan, Indonesia, Brazil, Saudi Arabia, and Morocco, engaged in a changing pattern of collaboration depending on specific needs.

The evolving geopolitical context accelerated the global networking strategy after the 2001 terrorist attack on New York. The legal and judicial constraints to state surveillance were significantly weakened in the name of national security and what was presented as a need to respond to the menace of Islamic terrorism. The Russian invasion of Ukraine, and the heightened tensions between the US and China, increased the power and the autonomy of the major surveillance agencies. Thus, the NSA is located in the Defense Department, unlike the CIA or FBI, and largely protected from congressional inquiries because of military imperatives. The British GCHQ, while formally dependent on the Foreign Secretary, is independent from the Foreign Office, as its director has the rank of Permanent Secretary.

The digital global surveillance system in the twenty-first century was greatly enhanced after the alarm triggered by terrorist networks. The NSA's director Michael Hayden declared that identifying potential threats at a global scale was like finding a needle in a haystack. And so he needed the entire haystack. President Bush and his successors issued several executive orders to circumvent the judicial controls that had been established. The War on Terror justified everything. The main judicial control, the authorization required by the Foreign Intelligence Security Act (FISA) court, remained. But in almost all instances, the court authorized the requests from the NSA. However, after 2001, the NSA realized that it was not on the cutting edge of digital surveillance technologies, so it embarked on a massive upgrading of its capability, modernizing some of its existing programs, and creating a flurry of new programs, thanks to technology transfer from the major US Internet companies, world leaders in digital technology.

Thanks to the revelations of whistleblower Edward Snowden,[1] we know the existence and characteristics of some of these programs. Particularly important was the PRISM program that facilitated access to the data of users of participating companies. Figure 3.1 shows the sequence of cooperation of the most important technology companies.[2] ICREACH is a search engine used by all US surveillance analysts to access billions of records of phone calls, emails, cell phone locations, and messages. UPSTREAM allowed the NSA to collect information on telecommunication traffic around the world by accessing intercontinental cables and Internet exchange points. TAILORED ACCESS OPERATOR was capable of retrieving two petabytes of data, the equivalent of three million CD-ROM discs, every hour. Other programs, increasingly secretive, include ROOM 641 in cooperation with AT&T, SPECIAL COLLECTION SERVICES, SENTRY EAGLE, MARINA, PINWALE, and MYSTIC (for voice interception). MUSCULAR, developed in cooperation with GCHQ, was specifically designed to intercept information in the networks of Google, prompting a strong public reaction from this company. BULLRUN was designed to break into encrypted networks, while OPTIC NERVE was used to collect webcam pictures from Yahoo! users. AURORAGOLD was used to intercept mobile phone calls and text messages. Other programs, such as QUANTUM, allowed the NSA to identify and replace the web content requested by a user. Meanwhile, XKEYSCORE was the in-house search engine that made it possible for NSA operatives to sieve through the vast amounts of data being aggregated by a variety of programs, making it possible for them to find details regarding specific individuals and institutions. XKEYSCORE was the way in which the needles were retrieved from the haystack. The Snowden

---

[1]   Edward Snowden is a former National Security Agency (NSA) contractor and whistleblower who released classified documents revealing the extent of the US government's surveillance programs in 2013. Snowden grew up in Maryland and enlisted in the US Army before beginning a career in the intelligence community. In 2013, while working for the NSA in Hawaii, Snowden shared classified documents with journalists, exposing a range of surveillance programs that he believed violated privacy rights and constitutional protections. He fled to Hong Kong and later Russia, where he was granted asylum since in the US he had been charged with espionage and theft of government property. Snowden's actions sparked a global outrage and debate on surveillance and privacy, which has led to significant reforms in government surveillance practices.

[2]   This slide was produced and used by the NSA for internal NSA briefings. It shows the companies actively cooperating with the PRISM program.

files inform on the basic characteristics of most of these programs. It goes without saying that other governments, particularly China, use similar technologies for their own spying and interception practices.

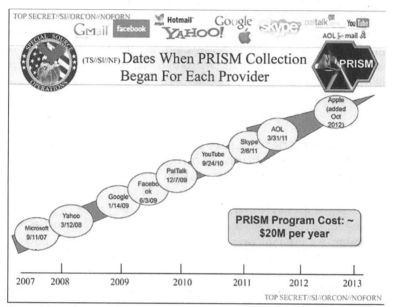

Source: Snowden—published by *Washington Post* (2013): https://da.wikipedia.org/wiki/Fil:Prism_slide_5.jpg.

Figure 3.1   Dates when collaboration with NSA began for each provider

The NSA collects so much data that it had to build a dedicated data center in Bluffdale, Utah. The data center covers one million square feet, roughly the size of 12 soccer fields, and is estimated to be able to store an exabyte or more of data, which would easily surpass the capacity needed to store voice recordings for all the phone calls made in the US in a year (Hill, 2013; Higgins, 2014).

However, if surveillance can operate at such proportions it is because of the systematic digitization of everything that both people do in their lives and organizations and institutions do in their daily management: personal interaction, credit cards, online banking, e-commerce, financial transactions, travel reservations, Uber or Lyft users' travels, online education, health services, real estate searches, search habits, entertainment (including adult websites, a major component of digital entertainment), cyber-religion, video-gaming, and the like. This is what Schneier (2016) calls the "digital exhaust," and what I would call "digital traces," whose formation and configuration have been analyzed by a number of experts in the field (Hepp et al., 2018). In short, Big Brother could not exist without the contribution of the Little Sisters that record the digital traces of our everyday life.

From Big Brother to Little Sisters: business models based on the collection, commodification, and processing of users' information

As the saying goes in Silicon Valley, "if you are not paying for a service, *you* are the product." This is clear in the case of major consumer-facing platforms. For instance, over 81% of the revenue of the largest Internet company, Alphabet, in 2021, came from advertising, particularly from targeted advertising enabled by the information that users provide through their searches on Google. The importance of advertising is even greater for Meta: in 2021 it accounted for almost 98% of its total revenue (Ang, 2022). Meta's strategies to extract revenue from data are diverse and some of them are company secrets. What we do know is that they go from direct ad placement on its websites to ad-related services, such as dashboards enabling customized aggregations that provide useful information on the social characteristics and geographical location of segments of users that can enable a corporate branding strategy, for example. In both cases, the amount and quality of information retrieved from users are at the core of their business model (Myers-West, 2019).

The business model of TikTok, at its launch in the US, was almost exclusively based on advertising. After 2021, services and technology added sources of revenue, such as in-app purchases, creator commissions, and branded hashtags. Much like in the case of other creator platforms, a key factor in the success of TikTok was the full cooperation of its users in editing, producing, remixing, and distributing videos across the app and beyond. This cooperation is mirrored by access to an audience of users powered by a black-box curation algorithm. Thus, the business model of

TikTok builds on an existing formula that combines advertising with free labor from the users (except for the minuscule percent that have amassed a large enough audience to get a creator commission from the platform). TikTok's novelty lies in the tools it incorporated into the platform to help users create content and the powerful algorithm that distributes content in a hyper-personalized way.

However, while retrieval of information from users for advertising purposes is at the core of some major Internet companies, particularly the platforms of social media networks, it is not the same for other Internet-based companies. In 2021, advertising was estimated to represent just over 1% of Apple's revenues.[3] For Microsoft, although the company does not disaggregate its total revenue in terms of advertising, we can estimate it falls somewhere between 5% and 15% (Microsoft, 2021; Ang, 2022). For Amazon, advertising represents merely 7% of revenues (Ang, 2022).

What is essential for all digital companies is the volume and quality of the traffic they generate in their networks, directly or indirectly. More specifically, what matters is the number of daily and monthly active users, frequency of their interaction, and quality of the information they generate directly and indirectly. These metrics are key in defining the financial valuation of these corporations, the most important source of capital accumulation for all companies.

In the case of Amazon, the information obtained from their users is essential for the optimization of the platform and targeted placement of their products. Moreover, information processing technologies are critical for a distribution company that is based on "click and brick," in which information systems underlie a vast transportation and warehousing network (Khan, 2017).

Thus, while commodification of users' data is essential in social media networks, it plays a lesser role in Internet-based companies at large. What has been discussed as *data capitalism* should be understood as just one segment of the digital economy that follows the systemic rules of informational capitalism, structured around the production and trade

---

[3] $4 billion out of a revenue of $365 billion in 2021 (Stokel-Walker, 2022; Statista, 2021).

of information and knowledge, and their application to all sectors of the digital economy (Peitz & Waldfogel, 2012; Cohen, 2019).

The connection between informational capitalism and the Surveillance State operates on the basis of the digitization of everything, on their storage of everything in databases, and on the circulation of information over multiple Internet-based communication networks. The technologies of surveillance used by the State focus on the interception of bytes circulating over these networks, be it social media, e-commerce, companies' interactions, or other communication agents. This is how what I call "Little Sisters" operate as the enabling factor for the growth of Big Brother.

The battle over privacy

At the 1999 launch of Jini, then a new program from Sun Microsystems, its CEO Scott McNealy made a statement that became a classic in Silicon Valley history: On the Internet "you have zero privacy anyway. Get over it" (Sprenger, 1999). However, people do not seem to have gotten over it. In the US, a Pew Research Center survey in 2014 found that 80% of users of social networking sites were concerned about advertisers or businesses having access to their data, and 70% were also concerned about government surveillance (Pew Research Center, 2014). Moreover, in other surveys users claimed that in reaction to news of surveillance they had changed their online habits to minimize sharing their data (Geiger, 2018). Similar concerns were expressed in the UK (Gibbs, 2015) and Europe at large (CIGI-Ipsos, 2019). As the massive breach of privacy by companies and governments became widely known, multipronged resistance came from various quarters.

A first line of resistance has come from a number of whistleblowers working in the security agencies, particularly the NSA, who sacrificed their careers and sometimes went to jail to denounce practices that were unlawful and endangered people's constitutional rights. The most notable was Snowden, who, knowing what had happened to his predecessors, took the precaution of leaving the country with key information files that were then published by mainstream media around the world. But there were a number of other whistleblowers from the NSA that came before him, such as William Binney (2002), Thomas Drake (2006), and Mark

Klein (2006). It is important to underline that whistleblowing on abuses of privacy went beyond the US and beyond the government. In the UK, for example, Christopher Wylie, from Cambridge Analytica, revealed the unauthorized use of Facebook data by his company to help the Brexit campaign in 2016, triggering a broad debate regarding the risks posed by the data stored under the control of social media platforms.

Julian Assange and Wikileaks were a key player in many revelations pertaining to illegal government practices. In 2017, for example, Wikileaks disclosed VAULT7, which compiled technologies and procedures used by the CIA for its hacking and cyberspying operations (Wikileaks, 2017). Assange took refuge in the Ecuadorian Embassy in London, and was soon after set to be extradited by the UK to stand trial in the US. The severity of the potential punishment on Assange and all of these ethical objectors, and the new security measures introduced by the agencies, aim to limit the string of revelations about these practices over time. Time will tell if they have been effective. In contrast, the European Union has rolled out legal protections for whistleblowers to ensure their agencies operate according to legal standards. Yet, the law has rarely been implemented, raising the question of whether potential whistleblowers actually trust these safety nets.

A second line of defense of privacy was built by techno-libertarians. These are skilled programmers who developed and distributed encryption, circumventing government surveillance. This clash—which placed government agencies and law enforcement on the side of weakening encryption, and the programmers (self-employed or working for large companies) on the side of promoting more widespread and powerful encryption—has been raging for a number of years now, and is colloquially referred to as *crypto-wars*. The father of the Internet, Vint Cerf, and the father of the World Wide Web, Tim Berners-Lee, are on record criticizing the excesses of the NSA and supporting advanced encryption to be used by citizens and companies as a barrier to unwarranted intrusion by surveillance agencies and unauthorized use of personal data (Ferenstein, 2013; Pilkington, 2013).

Some ethical hackers have leveraged military-funded programs and distributed them over the Internet. The most important case is The Onion Router (TOR), originally designed by the US Navy in 2002, and adapted by Roger Dingledine and Nick Mathewson to protect Internet traffic from government surveillance and operations aimed at repressing social movements. TOR helped the protesters considerably during the Arab Spring as well as other demonstrations around the world. Hacker Moxie Marlinspike developed the encrypted application Signal, a highly popular messaging app among activists and politicians around the world, and whose encryption technology has come to power mainstream applications like WhatsApp and Facebook Messenger.

Thirdly, in countries where civil society organizations are strong, they have taken companies and the executive branch to court. In the US, civil society organizations reacting to the revelations include the Electronic Frontier Foundation (EFF)[4] and the American Civil Liberties Union (ACLU),[5] and include Liberty in the UK (Siddique, 2021). In 2020, in a court case shepherded by the ACLU, the US Court of Appeals for the Ninth Circuit ruled that the warrantless telephone operation that secretly collects millions of Americans' telephone records violated the FISA and may well have been unconstitutional.

However, the conservative majority in the US Supreme Court reduced the protection of privacy that had come from legal actions from civil society organizations. Thus, in February 2023, the US Supreme Court denied the Wikimedia Foundation's petition for review of its legal challenge to the NSA's "Upstream" surveillance program. This came after, in 2022, the Supreme Court restricted women's right to an abortion in *Dobbs*, which put into question the hierarchy of the right to privacy within the US legal system (Gajda, 2022).

Governments in various countries came under pressure from their citizens, particularly in Europe, to provide stronger protections against government and business intrusion in their lives. Furthermore, they

---

4   In July 2013, EFF filed a lawsuit following revelations that an FISA court had ordered Verizon to turn over all customer phone records to the NSA, including who connected and for how long (EFF, n.d.).

5   In 2014 ACLU and others, including EFF, supported a lawsuit that challenged the government's bulk collection of telephone records, in particular questioning the notion that people have no expectation of privacy when they entrust information to others (EFF, n.d.). In 2016, ACLU obtained a victory when a Circuit Court found that the "staggering" amount of information collected by the NSA was a violation of the Fourth Amendment and the Patriot Act (Stempel, 2015).

protested to the US government regarding NSA spying practices that did not respect even leaders of allied countries, such as Germany's Chancellor Angela Merkel or Brazilian President Dilma Rousseff (Watts, 2013; MacAskill, 2015). It seemed that the dynamics of the surveillance regime were largely autonomous and not under close political supervision.

Internet companies became concerned about the reluctance of their users to accept the use of their personal data and about the growth of legal activism with potential judicial hostility to their intrusive practices. Furthermore, they realized the risks of cooperating too closely with government agencies that were penetrating their networks and databases with or without their consent. A renegotiation of the relationship between governments and Internet companies became necessary, impacting both the surveillance regime and the informational business model.

Resetting the cooperation between governments and information technology companies

Faced with increasing criticism from their customers about the depth of the intrusion into their lives, Internet companies took steps to show reform, including adopting technologies of encryption for the transfer and storage of user data, which could better protect users from third-party intrusions. In doing so, they were following the criteria put forward by most independent experts, including those at the World Wide Web Consortium (W3C) that stewards web standards, and of the Internet Engineering Task Force (IETF), which manages the Internet protocol, who had issued a statement asserting that "Well-implemented cryptography can be effective against pervasive monitoring and will benefit the Internet if used more" (Farrell, 2012), after which a coordinated effort toward secure Internet and web traffic was made.

Furthermore, digital rights non-profits and major companies like Facebook, Google, Twitter, and Reddit created a lobbying group named Reset the Net, to push back against pervasive government surveillance practices that were undermining the trust of their users. Google reacted to security agencies' intrusion into their own networks and storage facilities, issuing a formal statement:

> We have long been concerned about the possibility of this kind of snooping, which is why we have continued to extend encryption across more and more Google services and links … We are outraged at the lengths to which the government seems to have gone to intercept data from our private fiber networks, and it underscores the need for urgent reform (BBC, 2013).

The shift in public opinion that followed went further than what the companies could have expected. In 2018, thousands of Google employees publicly opposed collaborating with US military programs, requesting that the company discontinue Project Maven designed to update the artificial intelligence systems powering the image recognition capabilities in military drones (Statt, 2018).

This prompted a backlash from Republican congressman Matt Gaetz, who, in a hearing with big tech CEOs in 2020, asked:

> Do any of the rest of you take a different view? That is to say that your companies don't embrace American values. It's great to see that none of you do … Mr. Pichai is, did you weigh the input from your employees when making the decision to abandon [Project Maven] with the United States military?

Sundar Pichai, CEO of Google, answered: "Congressman, thanks for your concern. As I said earlier, we are deeply committed to supporting the military and the US government" (Rev, 2020).

As part of this hesitant reset, less scrupulous companies offered their services. As soon as Maven was abandoned by Google following employee resistance, other companies tried to move in and get this and similar contracts. For instance, Palantir, a government-focused company founded by Peter Thiel in 2003, came to the rescue of the Pentagon in 2019. It rebranded Project Maven as Project Tron, and moved forward the project of updating the AI capabilities of US military drones, signaling capacity and appetite to work on projects that most major multinationals considered unethical or reputationally risky (Greene, 2019).

The contradictory interests of tech companies came into the open. On the one hand, they have to preserve their lucrative contracts with the US government and, more important, to prevent being hit with anti-monopoly legislation. On the other hand, they could not lose the trust of their users in the US or risk their burgeoning businesses outside of the US for being perceived as tools of the US government (Ortiz Freuler, 2022), as Internet users in the US now represent just 7.5% of global users. And to do so they have to engage in the policy of encryption of their communication, making covert government surveillance operations more difficult, or at least that is the framing.

This tension between being a multinational company and growing geo-political tensions was also expressed by Microsoft's decision, in 2017, to try to bring some order to the networks of global collaboration by floating the idea of a Digital Geneva Convention on a global scale (Smith, 2017). However, it did not convince world leaders and was slowly dropped by Microsoft.

Yet, new frontiers of technological innovation have come to the forefront, opening new doors for digital surveillance. An example is Elon Musk's Starlink, largely financed by NASA, in an attempt to secure early dominance in low orbit space, where he is deploying a constellation of satellites that could soon replace traditional Earth telecommunications infrastructure. The rush is on to occupy these low orbit areas before Chinese (Chen, 2023) or European (Jones, 2023) corporations manage to step in, thus enabling the US to control the next centralized choke point through which governments might be able to pull and push data circulating across the network (Ortiz Freuler, 2022).

Elon Musk collaborated with Microsoft in investing in an innovative company OpenAI, which now offers services to the Microsoft suite of apps, including its search engine, Bing. In November 2022, OpenAI launched ChatGPT, a powerful AI chatbot application hosted on Microsoft servers, whose success made it possible for the company to gather a wealth of interaction data from its users, a key step in identifying what works and retraining the system, which contributed greatly to OpenAI's ability to continue to outpace its competitors. On the other hand, Google, Uber, Tesla, and others are in the race to develop AI technologies for self-driving cars, which includes deploying a wide variety of sensors on millions of vehicles that are constantly navigating the urban networks, thus broadening their capacity to collect information both from users and the urban environment.

Internet companies, particularly social media companies, restructured themselves seeking to diversify their business models. Over the past years they have been incorporating fees for sales in digital marketplaces (Constine, 2016), user subscriptions (Mehta, 2023), and sales of products like virtual reality headsets (Forbes, 2023) and smart glasses (Weatherbed, 2023), among a variety of strategies that could allow them to become less dependent on the commodification of personal information. This move is becoming urgent as users and regulators across the globe become more wary of the reutilization of these data for purposes that are not aligned with individual or national interests.

New actors are also coming in from what Silicon Valley used to perceive as a stagnant periphery: China. Outcompeted by TikTok among young users, Facebook rebranded as Meta and refocused resources toward the development of headsets that (they hope) will replace smartphones as the gates between the virtual and physical worlds. This enabled a public relations campaign that would refocus attention away from Facebook's poor performance and onto its promise of a future in which its billions of users migrate to the Metaverse, a place in which Facebook will not have to deal with Apple or Google, who currently control the market for operating systems on smartphones and have influenced the rules to ensure the prominence of their own products and data collection systems (O'Flaherty, 2022) (see Chapter 2). Meanwhile, Facebook and other US social media companies seem to be rallying Western governments to their rescue and to support them in undermining TikTok's growth by accusing it of spying for the Chinese government. Thus, the Snowden revelations created seismic waves that shook the old model of surveillance, sparked new business models, and reshuffled the actors. However, the need for surveillance, and their markets, never disappeared. They were merely transformed and innovated upon.

New surveillance technologies, new business models

Faced with resistance from civil society, judicial institutions, and new regulations to respond to pressure from citizens, government surveillance practices had to adapt. There was, however, a technical problem: companies increasingly encrypted their traffic against external snooping, while offering service to users in exchange for exploiting their personal data to an ever-increasing depth. With greater technical sophistication of encryption, and stricter regulation protecting privacy, the surveillance operations shifted to less detectable technologies able to circumvent encryption by shifting the attack to the edges of the network rather than the choke points in the middle. That is, while the companies had focused on effectively encrypting the information as it moved across the cables of the network, the intelligence agencies began targeting the smartphones, computers, and applications on the edges of the network, where they could access the content once it had been decrypted by each individual recipient.

Although there is now a flurry of this kind of surveillance technologies, I will exemplify the procedures with the widely exposed case of Pegasus. Pegasus is a malware program, developed by a private Israeli company, NSO (an acronym that stands for Niv, Shalev, and Omri, the names of the company's founders), for use by government surveillance agencies. The program infects a target's phone and sends back data, including photos, messages, and audio/video recordings to NSO clients. Pegasus's developer says that the software cannot be traced back to the government using it (NSO, n.d.)—a crucial feature for clandestine operations, although given the constant evolution of countersurveillance technology, I personally doubt such promises can be made.

However, as soon as a new surveillance technology appears, technologically savvy freedom fighters get to work. A University of Toronto-related, independent non-profit organization, Citizen Lab, dared to track the deployment of Pegasus. At the time of writing, they had found that 45 countries were affected by the espionage network. Files, photos, instant messaging, browsing history, location tracking, and social networks are the kinds of information that can be accessed.

A Catalan researcher working at Citizen Lab found evidence that the phones of Catalan pro-independence leaders have been hacked using Pegasus, a fact acknowledged by the Spanish Government. On the other hand, the phone of the Spanish Prime Minister and of the Defense Minister were also hacked, probably by Moroccan intelligence, and, separately, by Russian hackers during the Ukraine war.

Another notorious case was the intrusion into Jeff Bezos's smartphone. In this case, Pegasus was used for the retrieval of intimate and sexually explicit photographs with which he was later subjected to an attempt at extortion tied to his ownership of the *Washington Post*. Saudi Arabian intelligence has been suspected, with some tying the case to crown prince of Saudi Arabia, Mohammed bin Salman's discomfort with the *Washington Post*'s investigative reporting regarding the murder of its reporter, Khashoggi, at the hands of Saudi intelligence (Nast, 2020). Media reports claimed that the WhatsApp message that infected Bezos's phone was sent from the personal WhatsApp account of bin Salman. Even the richest person in the world could be exposed, at least for an instant.

The Bezos case prompted some governments, particularly in the US, to take action against NSO, although obviously the company denied responsibility. Thereafter WhatsApp decided to sue NSO for exploiting its infrastructure (Kirby, 2020). NSO claims it helps "government intelligence and law-enforcement agencies use technology to meet the challenges of encryption" during terrorism and criminal investigations. The company insists it works exclusively with government agencies (Priest et al., 2021), and that it cuts off access to Pegasus when it finds evidence of abuse by an agency (Kabir & Ravet, 2021). In its 2021 transparency report, the company claims it has exercised this power and cut off abusing clients in the past (NSO, 2021). However, an investigation released by Amnesty International in 2021 has shown that hundreds of journalists and activists that are unlikely to be terrorists have been targeted by the company, and in some cases, like in Mexico, seemingly linked to subsequent assassinations (Clark, 2021). As a consequence of this triad of forces, the US government eventually moved forward and blacklisted the NSO group, with catastrophic commercial consequences.

Meanwhile, the US contractor, L3Harris, a company that lists the Defense Department as its biggest government client, is said to have cited support from US intelligence officials in its effort to acquire the blacklisted NSO, and potentially wash its technologies and house them under a new company with ties to the US (Mazzetti & Bergman, 2022).

So, over these past decades, although surveillance technologies and mechanisms may change, critical information seems to always end up in the hands of the Global Surveillance Bureaucracy, while companies keep tracking the lives of their users for their profit-making strategies.

End of privacy?

The battle over privacy never ends. Living in a digital environment makes it more difficult than ever to preserve privacy. Yet, civil society mobilization, technologies of countersurveillance, citizen activism, informed journalism, judicial protection, and government legislation to respond to the concerns of their voters, show the persistence of the dialectics between power and counterpower as we enter a new regime of surveillance and new informational business models, increasingly dominated by information-hungry AI algorithms. The deep fears about machines (their algorithms, strictly speaking) taking over our lives are obscuring the actual contours of our digital existence: States and oligopolistic companies appropriate the new technologies without sharing information and understanding with citizens and clients. However, there will be no end to privacy as long as people's consciousness and critical thinking are ready and able to counter the techno-determinism that characterizes the new masters of the world.

References

Ang, C. (2022) "How do big tech giants make their billions?," *Visual Capitalist*. Available at: https://www.visualcapitalist.com/how-big-tech-makes-their-billions-2022/ (accessed on 7 March 2023).

BBC (2013) "Snowden leaks: Google 'outraged' at alleged NSA hacking," *BBC News*, 30 October. Available at: https://www.bbc.com/news/world-us-canada-24751821 (accessed on 3 March 2023).

Castells, M. (2000) *The Rise of the Network Society – The Information Age – Economy, Society, & Culture*. Malden, MA: Wiley-Blackwell.

Castells, M. (2001) *The Internet Galaxy: Reflections on the Internet, Business, and Society*. Oxford: Oxford University Press.

Castells, M. (2009) *Communication Power*. Oxford: Oxford University Press.

Castells, M. (2012) "Information technology and global capitalism," in W. Hutton & A. Giddens (eds), *On the Edge: Living with Global Capitalism*. New York: Random House, pp. 52–75.

Chen, S. (2023) "China to launch nearly 13,000 satellites to 'suppress' Starlink: researchers," *South China Morning Post*, 24 February. Available at: https://www.scmp.com/news/china/article/3211438/china-aims-launch-nearly-13000-satellites-suppress-elon-musks-starlink-researchers-say (accessed on 3 March 2023).

CIGI-Ipsos (2019) "CIGI-Ipsos global survey on Internet security and trust," *Centre for International Governance Innovation*. Available at: https://www.cigionline.org/cigi-ipsos-global-survey-internet-security-and-trust/ (accessed on 7 March 2023).

Clark, M. (2021) "Here's what we know about NSO's Pegasus spyware," *The Verge*, 23 July. Available at: https://www.theverge.com/22589942/nso-group-pegasus-project-amnesty-investigation-journalists-activists-targeted (accessed on 3 March 2023).

Cohen, J.E. (2019) *Between Truth and Power: The Legal Constructions of Informational Capitalism, Between Truth and Power*. Oxford: Oxford University Press.

Constine, J. (2016) "Facebook launches Marketplace, a friendlier Craigslist|TechCrunch," *TechCrunch*, 3 October. Available at: https://techcrunch.com/2016/10/03/facebook-marketplace-2/ (accessed on 3 March 2023).

EFF (n.d.) "NSA Spying," *Electronic Frontier Foundation*. Available at: https://www.eff.org/nsa-spying (accessed on 7 March 2023).

Farrell, H. (2012) "The consequences of the Internet for politics," *Annual Review of Political Science*, 15(1), pp. 35–52. https://doi.org/10.1146/annurev-polisci-030810-110815.

Ferenstein, G. (2013) "How the Internet's founders feel about the NSA scandal," *TechCrunch*, 31 December. Available at: https://techcrunch.com/2013/12/31/how-the-internets-founders-feel-about-the-nsa-scandal/ (accessed on 7 March 2023).

Forbes (2023) "VR headset sales underperform expectations, what does it mean for the metaverse in 2023?," *Forbes*, 6 January. Available at: https://www.forbes.com/sites/qai/2023/01/06/vr-headset-sales-underperform-expectations-what-does-it-mean-for-the-metaverse-in-2023/ (accessed on 19 October 2023).

Gajda, A. (2022) "How Dobbs threatens to torpedo privacy rights in the US," *Wired*, 29 June. Available at: https://www.wired.com/story/scotus-dobbs-roe-privacy-abortion/ (accessed on 7 March 2023).

Geiger, A.W. (2018) "How Americans have viewed government surveillance and privacy since Snowden leaks," *Pew Research Center*, 4 June. Available at: https://www.pewresearch.org/fact-tank/2018/06/04/how-americans-have-viewed-government-surveillance-and-privacy-since-snowden-leaks/ (accessed on 7 March 2023).

Gibbs, S. (2015) "Data protection concerns 72% of Britons in post-Snowden world, research shows," *The Guardian*, 9 April. Available at: https://www.theguardian.com/technology/2015/apr/09/data-protection-concerns-72-of-britons-in-post-snowden-world-research-shows (accessed on 7 March 2023).

Gill, P., & Phythian, M. (2018) *Intelligence in an Insecure World*. New York: John Wiley & Sons.

Greene, T. (2019) "Report: Palantir took over Project Maven, the military AI program too unethical for Google," *TNW|Artificial-Intelligence*, 11 December. Available at: https://thenextweb.com/news/report-palantir-took-over-project-maven-the-military-ai-program-too-unethical-for-google (accessed on 3 March 2023).

Greenwald, G. (2014) *No Place to Hide: Edward Snowden, the NSA, and the US Surveillance State*. New York: Metropolitan Books.

Hepp, A., Breiter, A., & Friemel, T.N. (2018) "Digital traces in context," *International Journal of Communication*, 12, pp. 439–449. https://doi.org/10.5167/UZH-148589.

Higgins, P. (2014) "Releasing a public domain image of the NSA's Utah data center," *Electronic Frontier Foundation*, 9 July. Available at: https://www.eff.org/deeplinks/2014/07/releasing-public-domain-image-nsas-utah-data-center (accessed on 25 February 2023).

Hill, K. (2013) "Blueprints of NSA's ridiculously expensive data center in Utah suggest it holds less info than thought," *Forbes*. Available at: https://www.forbes.com/sites/kashmirhill/2013/07/24/blueprints-of-nsa-data-center-in-utah-suggest-its-storage-capacity-is-less-impressive-than-thought/ (accessed on 25 February 2023).

Jones, A. (2023) "European Union to build its own satellite-Internet constellation," *Space.com*, 1 March. Available at: https://news.yahoo.com/european-union-build-own-satellite-140050794.html (accessed on 3 March 2023).

Kabir, O., & Ravet, H. (2021) "NSO CEO exclusively responds to allegations: 'The list of 50,000 phone numbers has nothing to do with us'," *CTECH*, 20 July. Available at: https://www.calcalistech.com/ctech/articles/0,7340,L-3912882,00.html (accessed on 3 March 2023).

Khan, L.M. (2017) "Amazon's antitrust paradox," *Yale Law Journal*, 126(3), pp. 710–805. https://www.yalelawjournal.org/note/amazons-antitrust-paradox.

Kirby, J. (2020) "The Saudi crown prince reportedly hacked Jeff Bezos," *Vox*, 21 January. Available at: https://www.vox.com/2020/1/21/21075990/saudi-arabia-crown-pince-mbs-amazon-jeff-bezos (accessed on 3 March 2023).

MacAskill, E. (2015) "Germany drops inquiry into claims NSA tapped Angela Merkel's phone," *The Guardian*, 12 June. Available at: https://www.theguardian.com/world/2015/jun/12/germany-drops-inquiry-into-claims-nsa-tapped-angela-merkels-phone (accessed on 7 March 2023).

Mazzetti, M., & Bergman, R. (2022) "Defense firm said US spies backed its bid for Pegasus spyware maker," *The New York Times*, 10 July. Available at: https://www.nytimes.com/2022/07/10/us/politics/defense-firm-said-us-spies-backed-its-bid-for-pegasus-spyware-maker.html (accessed on 3 March 2023).

Mehta, I. (2023) "Here's how every social media company is adopting subscriptions," *TechCrunch*, 27 February. Available at: https://techcrunch.com/2023/02/27/social-media-apps-adopting-subscription-models/ (accessed on 3 March 2023).

Microsoft (2021) *Microsoft 2021 Annual Report*. Available at: https://www.microsoft.com/investor/reports/ar21/index.html (accessed on 7 March 2023).

Myers-West, S. (2019) "Data capitalism: Redefining the logics of surveillance and privacy," *Business & Society*, 58(1), pp. 20–41. https://doi.org/10.1177/0007650317718185.

Nast, C. (2020) "If Saudi Arabia did hack Jeff Bezos, this is probably how it went down," *Wired UK*. Available at: https://www.wired.co.uk/article/jeff-bezos-phone-hack-mbs-saudi-arabia (accessed on 8 February 2023).

NSO (2021) *Transparency & Responsibility Report 2021*. Available at: https://www.nsogroup.com/wp-content/uploads/2021/06/ReportBooklet.pdf (accessed on 3 March 2023).

NSO (n.d.) *NSO Pegasus – DocumentCloud*. Available at: https://www.documentcloud.org/documents/4599753-NSO-Pegasus.html (accessed on 3 March 2023).

O'Flaherty, K. (2022) "Apple issues stunning new blow to Facebook as Google joins the battle," *Forbes*. Available at: https://www.forbes.com/sites/kateoflahertyuk/2022/02/19/apple-issues-stunning-new-blow-to-facebook-as-google-joins-the-battle/ (accessed on 3 March 2023).

Ortiz Freuler, J. (2022) "The weaponization of private corporate infrastructure: Internet fragmentation and coercive diplomacy in the 21st century," *Global Media and China*, 8(1), pp. 6–23. https://doi.org/10.1177/20594364221139729.

Peitz, M., & Waldfogel, J. (eds) (2012) *The Oxford Handbook of the Digital Economy*. Oxford: Oxford University Press.

Pew Research Center (2014) "Public perceptions of privacy and security in the post-Snowden era," *Pew Research Center: Internet, Science & Tech*, 12 November. Available at: https://www.pewresearch.org/internet/2014/11/12/public-privacy-perceptions/ (accessed on 7 March 2023).

Pilkington, E. (2013) "Tim Berners-Lee: Encryption cracking by spy agencies 'appalling and foolish'," *The Guardian*, 7 November. Available at: https://www.theguardian.com/world/2013/nov/06/tim-berners-lee-encryption-spy-agencies (accessed on 7 March 2023).

Priest, D., Timberg, C., & Mekhennet, S. (2021) "Private Israeli spyware used to hack cellphones of journalists, activists worldwide," *Washington Post*, 18 July. Available at: https://www.washingtonpost.com/investigations/interactive/2021/nso-spyware-pegasus-cellphones/ (accessed on 3 March 2023).

Rev (2020) "Tech CEOs face Congress at epic antitrust hearing," *Rev*. Available at: https://www.rev.com/transcript-editor/shared/_FX24Jlb75YkV0wn0tdgEzn7hr3YnKHiYFRaJHC36cpuN8-hRZCoC_eanIZkNRqAAoCUFtC5429mmv3rvjnTX3PpTLo?loadFrom=PastedDeeplink&ts=4121.04 (accessed on 14 December 2021).

Schneier, B. (2016) *Data and Goliath: The Hidden Battles to Collect Your Data and Control Your World*. Reprint edition. New York: W.W. Norton & Co.

Siddique, H. (2021) "GCHQ's mass data interception violated right to privacy, court rules," *The Guardian*, 25 May. Available at: https://www.theguardian.com/uk-news/2021/may/25/gchqs-mass-data-sharing-violated-right-to-privacy-court-rules (accessed on 7 March 2023).

Smith, B. (2017) "The need for a Digital Geneva Convention," *Microsoft on the Issues*, 14 February. Available at: https://www.microsoft.com/en-us/cybersecurity/blog-hub/need-digital-geneva-convention (accessed on 13 February 2023).

Sprenger, P. (1999) "Sun on privacy: 'Get over it'," *Wired*, 26 January. Available at: https://www.wired.com/1999/01/sun-on-privacy-get-over-it/ (accessed on 7 March 2023).

Statista (2021) "Global Apple ad revenue 2021," *Statista*. Available at: https://www.statista.com/statistics/1330127/apple-ad-revenue-worldwide/ (accessed on 7 March 2023).

Statt, N. (2018) "Google reportedly leaving Project Maven military AI program after 2019," *The Verge*, 1 June. Available at: https://www.theverge.com/2018/6/1/17418406/google-maven-drone-imagery-ai-contract-expire (accessed on 6 December 2021).

Stempel, J. (2015) "NSA's phone spying program ruled illegal by appeals court," *Reuters*, 7 May. Available at: https://www.reuters.com/article/us-usa-security-nsa-idUSKBN0NS1IN20150507 (accessed on 7 March 2023).

Stokel-Walker, C. (2022) "Apple is an ad company now," *Wired*, 20 October. Available at: https://www.wired.com/story/apple-is-an-ad-company-now/ (accessed on 7 March 2023).

Washington Post (2013) "NSA slides explain the PRISM data-collection program," *Washington Post*, 6 June. Available at: http://www.washingtonpost.com/wp-srv/special/politics/prism-collection-documents/ (accessed on 14 December 2021).

Watts, J. (2013) "Brazil to legislate on online civil rights following Snowden revelations," *The Guardian*, 1 November. Available at: https://www.theguardian.com/world/2013/nov/01/brazil-legislate-online-civil-rights-snowden (accessed on 7 March 2023).

Weatherbed, J. (2023) "Barely anyone is using Meta's Ray-Ban smart glasses," *The Verge*. Available at: https://www.theverge.com/2023/8/3/23818462/meta-ray-ban-stories-smart-glasses-retention-reality-labs (accessed on 19 October 2023).

Wikileaks (2017) *Vault 7: Projects*. Available at: https://wikileaks.org/vault7/#Pandemic (accessed on 24 February 2023).

4

The digitalization of financial markets: from derivatives to cryptocurrencies

The securitization of everything

Financial markets are the nervous system of the capitalist economy. The economic value of everything is determined by trade in the financial markets. Capital accumulation results from the valuation and devaluation of any assets by this trade. In order to be traded, assets become securities, that is, a fungible support of financial value that can be traded, such as currency, equity, debt, or derivatives. In the twenty-first century, the financial markets have been transformed by their global interdependence, and by the development of derivatives.

Derivatives are formalized by an over-the-counter contract between parties who agree to set a benchmark, or combination of assets, to establish the value of the derivative that they trade in a given time frame. Futures, options, and credit swaps are derivatives. Most of them constitute synthetic securities, because they combine the value, present or future, of their components, in a security that does not have a material existence, and whose value is assigned by trade in the market, without necessarily depending on the value of the underlying assets. This is because the derivative is simply the result of a mathematical formula, generally estimating probabilities of change in values, and because the market operates largely in terms of perception rather than according to rational estimates (Volcker, 2001: 75–85). As shown in Figure 4.1, the notional value of global derivatives was about six times higher than the global GDP in 2022.

Credit default swaps (CDS), a component of the derivatives market, reached a 5-year peak in transactions in 2023, with a total value of US$3,800 billion (Cooper, 2023). Although this is a small proportion

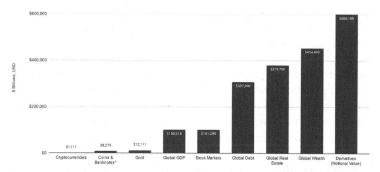

Source: Compiled and elaborated graphically by Juan Ortiz Freuler based on data from the following sources (ordered from left to right): CoinGecko (n.d.); BIS (2022 *Estimate for 20 major economies + EU); Companies Market Cap (2023); IMF (2023); SIFMA (2023); Bloomberg via IIF (2023a); Savills Global Research (2023 estimated); Credit Suisse (2023); BIS (2023).

Figure 4.1   Comparative analysis of global financial values (in US$ billion)

of the total derivatives market value, it is significant because the CDS, supposedly a mechanism to leverage risk, have been shown to be partly responsible for the financial crisis of 2008–10, as many of the loans that were swapped used unsecured mortgage loans as collateral, whose default brought down major financial companies, such as Lehman Brothers, ushering in the crisis. More significantly, they were at the source of the bankruptcy of AIG, the insurer of 50% of the banks in the world, then bailed out by the US Government to prevent an even greater collapse (Castells et al., 2012). In 2010, the US Securities and Exchange Commission (SEC) established new regulations concerning the CDS. Yet, their substantial increase as a proportion of credit derivatives being traded shows how difficult it is to regulate operations that are conducted in the markets with very little transparency, making proper accounting difficult because of the complexity of synthetic securities. This is because CDS, rather than a risk management mechanism, have been used as a strategy of speculation, betting on increase or decrease in values of a synthetic security whose composition depends, in a large number of eventualities, on the specificity of the multiple credits that are swapped. Financial regulators are aware of the risks involved in this market dynamics, but they also appreciate the growth potential offered by derivatives, inducing the creation of a gigantic reservoir of capital. While the value of this capital is largely virtual, it materializes in lending opportunities that keep the economy growing, albeit with increasing financial risk. Pity for the companies that default, particularly for the small ones (if they are too big to fail, they will be bailed out by the government—with taxpayers' money).

This transformation of financial markets has been made possible by widespread deregulation and liberalization of markets and financial institutions, and by the use of powerful mathematical models processed at high speed in computer networks. In the comparative research I led on the 2008–10 economic crisis, we showed that the complexity of the transactions, and the speed and volume of the exchanges, could only be managed by a flurry of new digital technologies (Castells et al., 2012). This trend was subsequently analyzed in depth by a group of researchers in Marszk and Lechman (2021). Given the deepening of the securitization of everything and the proliferation of the derivatives of endless securities, the coming of age of artificial intelligence was a most welcome development (Remolina & Gurrea-Martinez, 2023). However, while this new technological transformation provided better tools for calculations in private financial transactions, it did not improve the management capacity of the regulators. In fact, it enhanced the ability of the financial markets to strengthen their own unregulated dynamics. Capital returns resulting from the workings of a black box whose program was largely unknown favored the rise of what I conceptualized time ago as the "Financial Automaton." As I wrote: "Humankind's nightmare of seeing our machines taking control of our world seems on the edge of becoming reality, not in the form of robots eliminating our jobs or government computers that police our lives, but as an electronically based system of financial transactions" (Castells, 2001: 56). The Automaton is a network of digital networks that process all securities in the global financial markets on the basis of mathematical models without direct reference to their original underlying assets. The Financial Automaton fully materialized through high frequency trading (HFT) computerized systems, introduced in the New York Stock Exchange in 2002, increasing the speed of transactions to a 64 millionth of a second, compared with several seconds per transaction in the 1980s. In 2011, nano trading technology accelerated the speed, operating at 1 billionth of a second per transaction. Thus, the turnover of capital was boosted on the basis of digitalization of the transactions. Global capital is now largely virtual. Indeed, in terms of global wealth, in 2020 it was estimated that around 97% was held in digital form, and was traded over electronic networks (Mookerjee, 2021). This is why in 2023, the US Federal Reserve Board explored the possibility of creating a new currency, the "Central

Bank digital currency (CBDC)," backed by the monetary authority, equivalent in value to paper money. Although the Board did not dare to pursue this project at the time (they may eventually proceed), in fact they were late with their idea. Because in the meantime multiple tribes of techno-crypto libertarians were introducing into the financial markets a number of currencies that threatened to escape the control of the State and of financial intermediaries, thus potentially transforming the nature of money, of capital, and of capitalism. Not by ending it, but by expanding and deregulating its accumulation.

Cryptocurrencies

Economy is culture (Banet-Weiser & Castells, 2017) because transactions rely on a shared consensus of what is valuable. In each society, what is valuable depends on people's beliefs, and ultimately on the power relationships that make some values prevail over others. Capitalism is a culture enshrined in institutions as a result of power relationships that determine that the source of value is the capacity of capital to generate additional capital, measured in monetary units, as defined by the State. Currencies, as the formalization of money, are the communication means by which transactions can be enacted on the basis of the acceptance of a common standard of value that forms a transactional community (Swartz, 2020). This value is asserted by the issuer of the currency, usually the government, or a designated authority, and materialized in a technological support that provides the key practical features of any given currency: rarity (or managed scarcity), divisibility, and durability. The support required to assure these features evolved with technology. Gold was for a long time the support of a common standard for trading value, until it was replaced by paper money, which is a government-backed tender that would guarantee the nominal value, in its origin fictitiously equivalent to the amount of gold in government hands. Later came plastic money (i.e., credit/debit card accounts) issued by banks, denominated in legal currency, and supported by the assets and debts of cardholders. Thus, it is only fitting that in the digital society, a new form of money would emerge: a digital currency, that is a form of payment produced and operated in bits transacted over digital networks. Furthermore, since the digital society is simply the technological platform that allows the deployment of a new social structure—the network society (as I explain elsewhere in this volume)—the production of digital currencies denominated in a new measure of value, independent from governments and from the financial institutions, took place in networks designed to distribute the creation of value along peer-to-peer networks, bypassing any center of control. We must distinguish between digital trading of traditional currencies (operated in financial markets, as shown above), digital currencies, and cryptocurrencies, whose value is created, and traded in networks, using digital signatures that include users' secret private keys to verify their ownership of the assets against the records included on the decentralized ledger that is called a blockchain. Bitcoin is the most important of crypto-currencies, and so I will focus my analysis on it, although I will also refer to other currencies, like Ethereum, that expand the reach of the digital currency market.

The most significant feature of these currencies is that they are rooted in expressions of the crypto culture that emerged in the late twentieth century (Levy, 2001). The aims of the programmers, who called them-selves crypto-anarchists (1970s) and later cypherpunks (1990s), were distant from the financial world. They were primarily concerned with the defense of cryptographic technology to protect their privacy from the intrusion of the government and of financial companies. Max Brichta (2022, 2023) has provided a documented account of the cultural history of Bitcoin that clearly demonstrates how the technologies that gave birth to cryptocurrency networks emerged from successive waves of creative techies concerned about the concentration of power and capital result-ing, paradoxically, from the Internet revolution. Techno-rebels such as Ayn Rand and Timothy May, later joined by Eric Hughes in 1992, laid the ground for a number of projects aimed directly at the creation of a crypto digital currency as a means for securing people's autonomy. David Chaum's work, highly influential on May's and others', was more directly focused on transforming finance. Chaum also created DigiCash, a company designed to provide technology to make anonymous transac-tions without interference. Yet it was based on standard currencies and was subjected to the dangers of centralization of the network.

Several cypherpunks—most importantly Adam Back, the creator of the concept behind the "Proof of Work" algorithm that became central to the security of the Bitcoin network; Wei Dei, the designer of voluntary marketplaces, based on an expansion of online communities; and Nick Szabo, the programmer of BitGold, perhaps the first decentralized currency system—contributed the building blocks that made the creation and expansion of Bitcoin possible. This happened at the onset of the most major financial crisis in decades, on 31 October 2008. On that day, a message was posted on a mailing list for cryptographers created by Hughes and other crypto-hackers, signed with the pseudonym Satoshi Nakamoto.[1] The message started with the following sentence: "I've been working on a new electronic cash system that's fully peer-to-peer, with no trusted third party" (Franceschet, 2023). He went on to publish, in February 2009, a white paper "Bitcoin: A peer-to-peer electronic cash system," that was commented on and improved by the network of cryptographers, in particular with regard to software security. The system was based on blockchain technology that later migrated from this application to a large number of applications used in multiple domains. Without going into the specification of blockchain technology, which can be easily consulted in multiple sources (e.g., Baucherel, 2020), I will simply emphasize the features that result from an application of the cypherpunk culture into networked digital technology. Two characteristics are essential: first, a networked public ledger that is protected by cryptography; second, a mechanism to record and store data regarding the transactions so that the computers on the network can authenticate each transaction according to a "Proof of Work" mechanism, that is, a record that provides proof of when the transaction has happened, and which requires high computation power in order to disincentivize rogue actors from attempting to manipulate the ledger. Each transaction is consecutively appended to an ever-growing ledger of chained transactions, meaning it is incredibly transparent (albeit pseudonymous). In fact, given this never-ending chain of consecutive transactions, the technology should have been labeled *time*-chain. Yet, blockchain has become the popular name of this digital

---

[1]   Satoshi Nakamoto does not exist under such a name. It may be a person or a group of networked persons. He/she/it has never been identified, and the claims of a few people to be Satoshi Nakamoto have been dismissed upon further investigation. It is probably someone or some people educated in the UK because of the English spelling. We know "Satoshi" is the creator of Bitcoin because it originated the first transaction in the blockchain network, although the recipient of the transaction may have been himself. The last trace dates to 2011 when it communicated to someone in the network that "I have moved to other things." Behind its mystery it may well be that, as one of the goals of cypherpunks, it was trying to avoid taxation. Because it was the original miner, its fortune in Bitcoins may have reached US$78 billion at the height of the Bitcoin frenzy in 2021.

network technology. In a blockchain, the consensus mechanism, involving hundreds of distributed computers, makes it practically impossible to delete or alter any past records, so that, in its application in Bitcoin, the distributed public ledger records the transaction and also calculates its value in Bitcoins.

The production of new Bitcoins takes place by "mining." The Bitcoin code limits the number of Bitcoins that can be minted to a maximum of 21 million, thus ensuring the scarcity of the currency to limit its devaluation, or at least this was the goal. Mining involves running the Bitcoin code to verify transactions, and solving a given "hash," that is a mathematical problem whose solution requires a certain amount of computational power that grows with the number of transactions. The miner that first solves the hash is rewarded in Bitcoins in an amount that halves roughly every 4 years, another key deflationary characteristic of the design. Thus, whereas the reward for successful miners was 12.5 Bitcoins until 2020 (around US$45,000 in February 2019 prices but US$147,500 at August 2020 prices), it dropped to 6.25 Bitcoins in 2020–23 (roughly worth US$172,161 in October 2023), and at some point in 2024, it will become 3.125 Bitcoins. After this reward is assigned, the block of transactions is added to the blockchain with copies being verified by thousands of computers across the network. The process restarts, with a different hash, approximately every 10 minutes. Once a Bitcoin has been generated through the mining process, its owner can trade it for other cryptocurrencies or standard currencies on digital exchanges, thus contributing to price dynamic.

However, to fully understand the nature of Bitcoin, I must emphasize—as pointed out by Brichta—that it is both a process in a network and an asset as the product resulting from the process. Without the network, running on the basis of a precise code, there would be no Bitcoins, and so Bitcoin is a network creation, with the network formed by miners who share certain values and own significant computing capacity. Once Bitcoins are created, they are financial assets, traded in the financial market. This double characteristic induces the misunderstanding of Bitcoin as a purely speculative system. In fact, Bitcoin *traders* speculate in Bitcoins just as financial *traders* in general do on every traded asset, as I described in my analysis of securitization in the financial markets. Indeed, securitization is another form of digitizing the trading of value, since most securities now are synthetic derivatives from practically untraceable origins, while the logic of the miners, still rooted in the culture of cryptographic freedom, can be better characterized by pursuing accumulation of value in the network and in the individual nodes managed by the participating miner. For most of them, mining Bitcoins is a form of communication, or speech, and even free speech. As Wales and Ovelmen (2019: 210) argue, "Bitcoin is an ideological technology that was created specifically to allow its users to associate with a broad global network of individuals who share common values through a communication network that rejects the need to depend on centralized intermediaries" (cited by Brichta, 2022).

In 2023 it was calculated that there were around 12 million monthly transacting wallets that were moving roughly US$100 billion each month, with over 100 million wallets transacting yearly, and 50,000 developers were working regularly on crypto-related GitHub repositories (Horowitz, 2023).

The scarcity built into the Bitcoin algorithm, and the increasing insecurity in financial markets, has driven a substantial increase in its valuation (and of other cryptocurrencies) over the last decade. Between 2017 and 2023, for example, the value of Bitcoin saw an increase of 2,562%, while the Dow Jones Industrial Average grew by 75.2% and the US Dollar Index 9.97%. On the other hand, crypto markets are characterized by volatility, and by some spectacular crashes, as in 2020–21. However, at the time of writing, Bitcoin seems to be substantially less volatile than the main stock exchange indexes (Bloomberg, 2023b). Regardless of the perception of volatility, the market capitalization of digital currency markets, including crypto, remained high in 2023, reaching an estimated value of over US$1 trillion.

The growing public attention on cryptocurrencies has prompted major banks, credit companies, and financial institutions to seek a piece of the Bitcoin trading markets. Banks such as Citi, Bank of America, Bank of New York, and others, as well as financial companies as important as Goldman Sachs, have either announced interest or are already offering institutional investors and valuable clients the ability to trade in crypto-currency through their accounts.

Thus, it would appear that digital currencies are here to stay, in a striking corroboration of the fact that culture creates value, even though any value created in a capitalist economy is also added to the mother of all valuations, the global financial markets.

Blockchain platforms, digital currencies, and smart contracts

The technology underlying Bitcoin, blockchain, has become widely used as support for a number of new forms of digital financial transactions. A particularly important type of transaction is the so-called "smart contract." Through them, contracting parties inscribe the contractual conditions on the blockchain, allowing for the transfer of assets to take place automatically once the meeting of conditions is certified by the network, without needing the intervention of any other third party or authentication organization. This is tantamount to automating transactions and therefore it reduces transaction costs and increases the velocity of the turnover of capital involved in the transaction. The design of the original Bitcoin blockchain was adapted by Ethereum and others to better serve this type of service. Thus, Ethereum enabled a flurry of new digital currencies and tokens to enter the market, on the basis of a technology that, in addition to offering the characteristics of a cryptographic digital currency, also facilitated "smart contracts" and similar operations. Ethereum was introduced in 2014, and its blockchain supports its own currency, the "ether," as well as a variety of distributed applications, including "smart contracts." Ethereum's continued success is also due to its flexibility. In 2022 it successfully reformed a core element of the original blockchain protocol with the purpose to achieve a more efficient management of the energy involved in the consensus mechanism. As a result, Ethereum energy consumption dropped by 99% and it now uses 30,000 to 50,000 times less energy than Bitcoin (Horowitz, 2023), making it a key sticking point for individuals and companies seeking to have a greener portfolio.

Blockchain technology has also supported the development of smart contracts that made it possible to issue "digital tokens," each one of the tokens following its own rules, according to the code that runs the network. Thus, tokens are often issued as a vehicle for the funding of new projects or to otherwise assign roles and responsibilities to members, which, according to some analysts, leads to a "token digital community" with expectations within such communities of escalation toward a "token economy." Tokens can typically be traded, and so eventually they are also incorporated onto the financial market.

To allow people to keep their assets within their digital wallets in times of volatility, "stable coins" were introduced. These are digital currencies whose value is pegged to a financial asset or to a standard currency, like the US dollar. Many of the biggest scandals came from companies that claimed to offer a stable coin, but did not have the corresponding reserves, triggering investigations from the SEC that they are selling an unregistered security (Forbes, 2023). When a digital currency is backed by a central bank it becomes a CBDC that, while being digital, is not necessarily nor typically blockchain based, but rather a virtual representation of paper money, known and regulated by the financial authorities.

The empire strikes back: the State and cryptocurrencies

Cryptocurrencies emerged as an expression of a libertarian culture trying to escape the control of governments over transactions between people, and their subsequent taxation, by using their own self-generated currency behind the safety of encryption and software-managed network interaction. This is unacceptable for the State. Because, as Cory Doctorow reminds us, the historical origin of money is to be a facilitator of transactions of goods and services among people that had to be rendered visible, so they could be taxed (Doctorow, 2020). Therefore, it is logical that the Chinese State, founded on the principle of exercising effective control over the economy and society, would be the most determined opponent of Bitcoin and other cryptocurrencies, particularly because 26% of the global mining took place in China in 2020. Therefore, in 2021, the Chinese Government banned all trading and mining of Bitcoin. Many of the miners were forced to emigrate to continue their trade in more hospitable lands. However, some mining seems to continue under cover. The Chinese Government has also been tightening up its control over all forms of unregulated digital finance. For instance, in 2020 it blocked the initial public offering (IPO) of Ant, which was expected to be the largest IPO in history (Canales, 2020). Ant, a fintech company that is part of the Alibaba Group—the most valuable financial company in China—was trying to become a lending company beyond the control of regulators.

The company's CEO, the legendary Mr. Ma, was forced into informal exile in Japan.

At the same time, the Bank of China has been piloting its own CBDC, the e-CNY or digital yuan, which will provide the government with real-time insights regarding the country's economic heartbeat, as well as a tighter control over transactions. At the time of writing the pilot already had over 100 million wallets and over US$250 billion in transactions, although it represented only 0.16% of the circulating money supply (Wee, 2023). The US Federal Reserve Board, as mentioned above, was exploring the creation of a similar currency so as to not be left behind in the global race toward new money, and it may eventually introduce it. Indeed, by June 2023, 11 countries (mostly from the Caribbean) had launched their CBDCs, while another 78 were in the process of researching before proceeding. In the US, the main difficulties appeared to be the fear of the Federal Reserve that the introduction of a stable digital currency backed by the government would trigger a rush of banking customers to transfer their savings into more convenient means of payment, especially among the digital natives, thus hurting bank deposits.

Two small states, El Salvador and the Central African Republic, have already tried to join the digital trend by recognizing Bitcoin as legal tender in their country. However, the low level of digital literacy among their population, and the incapacity of regulators to ride the tiger of the global cryptocurrency markets, at least at the time of writing, seem to have provoked considerable losses to their economies and to the government's budget (Ongweso, 2022).

Techno-libertarians were able to reinvent money, as digital currencies are expanding at a fast rate, thus changing financial markets. Yet, while the original intent was to move into cryptocurrencies that would free people's money from the State, they've had to face the Leviathan that has haunted human history. To be continued.

References

Banet-Weiser, S., & Castells, M. (2017) "Economy is culture," in M. Castells (ed.), *Another Economy Is Possible: Culture and Economy in a Time of Crisis.* Cambridge: Polity, pp. 4–33.

Baucherel, K. (2020) *Blockchain Hurricane: Origins, Applications, and Future of Blockchain and Cryptocurrency.* New York: Business Expert Press.

BIS (2022) "How much money is in the world right now?," *GOBankingRates.* Available at: https://www.gobankingrates.com/money/economy/how-much-money-is-in-the-world/ (accessed on 13 October 2023).

BIS (2023) "OTC derivatives statistics at end-December 2022." Bank for International Settlements. Available at: https://www.bis.org/publ/otc_hy2305.htm (accessed on 13 October 2023).

Bloomberg (2023a) "World debt surges to record $307 trillion, rises as share of GDP," *Bloomberg.com*, 19 September. Available at: https://www.bloomberg.com/news/articles/2023-09-19/world-debt-surges-to-record-307-trillion-rises-as-share-of-gdp (accessed on 13 October 2023).

Bloomberg (2023b) "Bitcoin turns less volatile than S&P 500, tech stocks and gold," *Bloomberg.com*, 1 August. Available at: https://www.bloomberg.com/news/articles/2023-08-01/Bitcoin-turns-less-volatile-than-s-p-500-tech-stocks-and-gold (accessed on 6 October 2023).

Brichta, M. (2022) "Bitcoin, the network. An analytic history of Bitcoin's origins and affordances," Research Paper for Doctoral Seminar Comm 647, Los Angeles, USC Annenberg School of Communication.

Brichta, M. (2023) "Bitcoin as critique: On the socio-technical foundations of Bitcoin maximalism," Research Paper for Doctoral Seminar Comm 670, Los Angeles, USC Annenberg School of Communication.

Canales, K. (2020) "Jack Ma hasn't been seen in public since Ant Group's IPO was pulled. Here's how Chinese regulators slammed the brakes on the firm's would-be record-breaking $37 billion IPO," *Business Insider*, November. Available at: https://www.businessinsider.com/what-happened-ant-group-ipo-jack-ma-alipay-2020-11 (accessed on 5 October 2023).

Castells, M. (2001) "Information technology and global capitalism," in W. Hutton & A. Giddens (eds), *Global Capitalism*. New York: The New Press.

Castells, M., Caraça, J.M.G., & Cardoso, G. (eds) (2012) *Aftermath: The Cultures of the Economic Crisis.* Oxford: Oxford University Press.

CoinGecko (n.d.) "Crypto market cap charts," *CoinGecko*. Available at: https://www.coingecko.com/en/global-charts (accessed on 13 October 2023).

Companies Market Cap (2023) *Market Cap of Gold (precious metal)*, 4 October. Available at: https://companiesmarketcap.com/gold/marketcap/ (accessed on 13 October 2023).

Cooper, A. (2023) "Explainer: What are credit default swaps and why are they causing trouble for Europe's banks?," *Reuters*, 30 March, sec. Markets. Available at: https://www.reuters.com/markets/what-are-credit-default-swaps-why-are-they-causing-trouble-europes-banks-2023-03-28/.

Credit Suisse (2023) *Global Wealth Report.* Available at: https://www.credit-suisse.com/about-us/en/reports-research/global-wealth-report.html (accessed on 13 October 2023).

Doctorow, C. (2020) "Ant, Uber, and the true nature of money," *Pluralistic*, 5 November. Available at: https://pluralistic.net/2020/11/05/gotta-be-a-pony-under-there/ (accessed on 5 October 2023).

Forbes (2023) "The SEC's stablecoin crackdown could reshape the entire crypto market," *Forbes*, 21 February. Available at: https://www.forbes.com/sites/qai/

2023/02/21/the-secs-stablecoin-crackdown-could-reshape-the-entire-crypto
-market/ (accessed on 5 October 2023).

Franceschet, M. (2023) "Decentraland: The alleged decentralization of blockchain applications," *Communications of the ACM*, 66(6), pp. 45–47. https://doi.org/10.1145/3563942.

Horowitz, A. (2023) State of Crypto. Available at: https://api.a16zcrypto.com/wp-content/uploads/2023/04/State-of-Crypto.pdf (accessed on 14 September 2023).

IMF (2023) "Global GDP 1985–2028," *Statista*. Available at: https://www.statista.com/statistics/268750/global-gross-domestic-product-gdp/ (accessed on 13 October 2023).

Levy, S. (2001) *Crypto: How the Code Rebels Beat the Government – Saving Privacy in the Digital Age*. London: Penguin.

Marszk, A., & Lechman, E. (eds) (2021) *The Digitalization of Financial Markets: The Socioeconomic Impact of Financial Technologies*. London: Routledge.

Mookerjee, A.S. (2021) "What if central banks issued digital currency?," *Harvard Business Review*, 15 October. Available at: https://hbr.org/2021/10/what-if-central-banks-issued-digital-currency (accessed on 13 October 2023).

Ongweso, E. (2022) "The Bitcoin crash is taking El Salvador's big bet with it," *Vice*, 14 June. Available at: https://www.vice.com/en/article/n7zwj7/the-Bitcoin-crash-is-taking-el-salvadors-big-bet-with-it (accessed on 5 October 2023).

Remolina, N., & Gurrea-Martinez, A. (2023) *Artificial Intelligence in Finance: Challenges, Opportunities and Regulatory Developments*. Cheltenham, UK and Northampton, MA, USA: Edward Elgar Publishing.

Savills (2023) "Total global value of real estate estimated at $379.7 trillion – almost four times the value of global GDP," *Savills News*. Available at: http://www.savills.de/news---assets/article_hidden.aspx?id=352068 (accessed on 13 October 2023).

SIFMA (2023) *The Capital Markets Fact Book*. Available at: https://www.sifma.org/resources/research/fact-book/ (accessed on 13 October 2023).

Swartz, L. (2020) *New Money: How Payment Became Social Media*. New Haven, CT: Yale University Press.

Volcker, P. (2001) "The sea of global finance," in W. Hutton & A. Giddens (eds), *Global Capitalism*. New York: The New Press.

Wales, J.S., & Ovelmen, R.J. (2019) "Bitcoin is speech: Notes toward developing the conceptual contours of its protection under the First Amendment," *University of Miami Law Review*, 74(1), pp. 204–275.

Wee, R. (2023) "China's digital yuan transactions seeing strong momentum, says cbank gov Yi," *Reuters*, 19 July. Available at: https://www.reuters.com/markets/asia/chinas-digital-yuan-transactions-seeing-strong-momentum-says-cbank-gov-yi-2023-07-19/ (accessed on 5 October 2023).

5 Teleworking and the networked metropolis

Urbanization and digital communication

For a long time, futurologists were predicting that the advent of digital information and communication technologies (ICTs) would allow for widespread teleworking, so decreasing the need for spatial concentration of population and activities. It would follow a pattern of territorial sprawl that would contain urban growth. Instead, while connectivity, digitization, and Internet use have expanded at an accelerated rate, in the last four decades the planet has experienced the largest wave of urbanization in human history. The percentage of urban population in the world was 39% in 1980, 47% in 2000, 57% in 2023, and is projected to reach 68% by 2050. Moreover, this process of urbanization is characterized by the increasing relevance of a new spatial form that I conceptualized as the metropolitan region (Castells, 2010). Although size of population is not the only distinctive feature of the metropolitan region, using it as an indicator shows the trend: in 1960 there were only two metropolitan areas with more than ten million inhabitants, while in 2022, there are 44. The reasons for this apparent paradox are well known to urban research. Population follows jobs, and jobs result from economic activity, both in production and consumption. Spatial concentration favors economies of scale in infrastructure, labor resources, ancillary services, and consumption markets. Furthermore, in the knowledge and information economy that characterizes the Information Age, economies of synergy became paramount in fostering innovation and productivity. Synergy is favored by recurrent interaction among knowledge producers. Interaction is a function of the extent of communication bandwidth, and the broadest bandwidth is direct human communication. Decision making in all human activities requires multiple dimensions of communication that cannot be fully performed at distance, such as confidentiality, personal trust, and synchronous cooperation. The spatial concentration of high-level management functions explains the persistence of concentration in central business districts (CBD). The synergy of direct interaction explains the role of territorial concentration in the formation of milieux of innovation, such as technology areas, artistic hubs, or research universities (Castells & Hall, 1994; Hall, 1998). Besides, history and geography played a key role in keeping in place the upper layer of decision-making and innovation centers. The real estate value of buildings and facilities in these locations would be lost by moving out, unless the whole production and innovation complex were to move at the same time. This is why London's City, New York's Wall Street, Silicon Valley, or the Harvard-MIT campuses remain where they are even if they enjoy the most powerful telecommunication facilities in the world that could make possible their decentralization.

Since the high value producers live and work in certain areas, labor pools, consumer services, and the markets that result from this concentration keep attracting population and increasing real estate prices, feeding metropolitan concentration as an expression of the concentration of wealth, power, and knowledge. However, a large share of economic activities and their labor needs do not require the same level of economies of synergy. As for economies of scale it depends on the relationship between the productivity that can be achieved by work at distance and the price of real estate. If work at distance does not affect productivity excessively, the cost-saving strategy of leaving the main metropolitan centers and/or dispersing labor could be attractive to employers. This follows substantial intra-metropolitan decentralization with multiple secondary centers of information production, and a residential sprawl that is largely dependent on housing affordability and the life cycle of the residents. Therefore, there are two processes shaping the spatial structure at the same time: metropolitan concentration and intra-metropolitan decentralization.

The spatial result of this process is the metropolitan region that is connected to other metropolitan regions in the country and in the world by telecommunication and information systems. Because they are the centers of accumulation of capital, talent, and technology, and subsequently privileged spaces for urban amenities, they increasingly attract high-level functions, skilled workers, entrepreneurs and innovators, and feed themselves with global exchanges, forming a global network of metropolitan regions (Taylor & Derudder, 2016). The intra-metropolitan space is highly diversified, encompassing built areas, open space, and in some cases agricultural production. The metropolitan areas are internally networked by multimodal transportation networks and Internet connections that allow for home-based entertainment, and delivery of shopping and food staples. This is the networked metropolis, that is the spatial patterning of the new urban form of the Information Age, the metropolitan region. This spatial pattern emerged gradually *without significant increase of teleworking.* Functional activities were networked, households were networked (physically and digitally), global and metropolitan transportation and Internet networks connected economy, culture, and society around the planet. But most people, including most knowledge producers, continued to be submitted to exhausting, life-consuming daily commuting, contributing to congestion and pollution, ultimately adding to our self-destructive process of climate change. Then Covid-19 struck in 2020. And we were forced to protect ourselves from bodily contact by migrating everything that could be migrated to digital networks over the Internet. For a while, teleworking exploded everywhere, and we discovered that we could function in a new world of teleworking and tele-everything, waiting for the salvation of vaccines produced by science. The question arises: does this development signal a paradigm shift in organizing work, life, and the city?

Teleworking: a retrospective perspective

Teleworking rarely refers to exclusively working from home. In fact, many professionals sometimes work from transportation, cafes, libraries, and the like (Castells et al., 2006). Thus, I will refer to "remote work" or "non-workplace working location." It is usually considered "remote work" if someone works from home half of the working days in a 4-week time span.

Looking back, the pop notion of the "electronic cottage" (Huws, 1991) never materialized. In the US, some scholars predicted in the 1980s that, by 2000, 20 million people would be teleworking. In fact, at that date only 4.2 million (3.3% of the labor force) were "usually" working from home (Mateyka et al., 2012). By 2019, just before the pandemic, they remained at 4.7% of the labor force (Barrero et al., 2021).

In the European Union in 1999, only 2% of the labor force worked from home at least once a week (Gareis & Kordey, 2000), with Finland being at the top (6.7%) and Spain being at the bottom (1.3%) In the UK, just about

2.4% were teleworking under this broad definition. In 2019, in the EU, only 3.2% were "usually" working from home (Eurofound, 2022).

The Covid-19 pandemic in 2020 and 2021 changed drastically the use of teleworking. The study by Brynjolfsson et al. (2020) showed that, in the US, 35.2% changed their activity to teleworking, adding to the 15% that were already working remotely some of the time. Younger workers were the most likely to switch to remote working. Women favor working at home slightly more than men. Areas with higher proportions of management and professional employees experienced a higher increase in teleworking. Indeed, according to the study by Barrero et al. (2021), information workers were working from home 2.28 days per week, while manufacturing workers did so for only 0.93 days. Hansen et al. (2023) showed a particularly strong increase in remote work jobs posting between 2019 and 2023 in San Francisco (+491%), Boston (+437%), and New York (+420%). The same study showed a similar disparity between cities in the UK, with strong concentration in the main cities such as London (+693%) and Edinburgh (+811%). In the European Union, the share of employees working from home some of the time went from 11.1% in 2019 to 21.9% in 2021. Similar trends were found in a study of 20 OECD countries (Adrjan et al., 2022), with the exception of Japan where the number of teleworkers remained stable. In general, the increase in remote work came from workers who were working in-person before. This study included a survey of workers' feelings on their experience of teleworking. The majority considered it a positive experience. Fifty-four percent said they had higher productivity in contrast to only 13% who felt less productive. They perceived major benefits for their lives in terms of saving time, flexibility in organizing work, fewer meetings, and avoiding long commuting. In the US, 61 million minutes per day were saved by avoiding commuting during the pandemic. Women valued benefits from teleworking more than men. In a survey by Aksoy et al. (2022), 25% of respondents said they would quit their job if they were compelled to return to the office. They were ready to reduce their earnings up to 5% to pay for remote work 2 or 3 days per week. In the US, 33% said they would leave their job if they had to return full time to the office. This attitude was stronger for highly skilled professionals, those most coveted by companies.

Managers are positive about teleworking in general, but less than workers because they fear loss of control of the work process (Criscuolo et al.,

2021). However, both workers and managers expressed their preference for a hybrid model, spending some time in the workplace and some time working at home. This is because they also need to socialize with colleagues and they need to maintain space-time boundaries between work and home life. In fact, in the US, hybrid working represents around 25.6% of the workforce, while full remote work is limited to 7.9%. This sudden increase in teleworking was favored by technological innovation that provided new tools for collaborative working at distance, such as Zoom, Microsoft Teams, and Webex (Florida et al., 2023). The fast progress in artificial intelligence is enhancing productivity in processes of remote work thanks to improved human–AI interaction, and automation of communication (Jarrahi et al., 2023).

However, the jury is still out concerning the intensity of teleworking after the partial confinement of the Covid-19 pandemic. Most activities returned to the usual pattern, particularly in education, health, government, and social services. Yet, the accumulated experience was used in increasing distant working cooperation, for instance in education. In the private sector, the information-intensive activities kept increasing the frequency of teleworking, to save office space and location costs in expensive areas, and to make life easier for workers. The persistence of teleworking varied substantially depending on sectors and on geographic areas, depending on the relative importance of information processing employment in their labor pools.

At the time of writing (2023), studies that evaluated teleworking in the post-pandemic situation were scarce. However, an interesting study by Hansen et al. (2023) projected growth of teleworking by analyzing remote work job postings. They appeared to be increasing in a number of countries. In the US, the number of job offers explicitly mentioning remote work increased from 3.5% in 2014 to 12.5% in January 2023. The study was able to evaluate vacancy postings that explicitly offered hybrid or fully remote work from 2020 to 2023 in five industrialized countries.

In the US, the share of postings of remote work offerings increased substantially in San Francisco, Boston, and New York and stabilized in 2022–23, mainly reflecting the growth of information and knowledge sectors. Since these are the sectors that lead economic growth and add highly skilled employment, it would not be fanciful to project the increase of remote work or hybrid work in the coming years.

An emerging spatial structure?

Concentration of activities and residence in the largest metropolitan areas seems to have slowed down in the post-Covid years, while still being the predominant form of human settlements around the world. An insightful analysis by Marley Randazzo using US Census data (Randazzo, 2022) shows a large metropolitan outmigration in the 15 most populated metropolitan statistical areas (MSAs) in 2021–22, with San Francisco and New York leading the trend. There is a flow of residents into smaller metropolitan areas and some rural counties. However, at the same time, international immigration concentrates in the largest metropolitan areas, where job opportunities are the highest, compensating some of the loss of population in these cities. What is a definite trend is the outmigration out of the urban core, following the pattern of metropolitan decentralization into suburbs and exurbs. Ramani and Bloom (2021) provided additional evidence on the decentralization pattern. They showed that in 2021 the densest urban areas lost 9% of their population and 16% of their businesses relative to pre-Covid-19 levels. Fifty-eight percent of the movers relocated within the same metropolitan areas, into suburbs and exurbs, both for population and business, while 29% moved to midsized metro areas, 9% went to other large cities, and 4% to rural areas. While telework facilitates this movement, the main reasons for this outmigration seem to be the increase in real estate prices and the life cycle of the large cohort of millennials. The urban renaissance based on the appeal of the culture of city life diminishes as millennials reach the age of forming families rather than frequenting bars (Myers, 2016). The possibility of teleworking may be more relevant to explain the relocation of offices outside the CBDs, influenced by the impact of rising real estate costs as a result of the concentration of valuable activities in a limited space. Traffic congestion, pollution, and the perception of unsafe environment are factors related to employee preference for the hybrid model of work location, including as much the wishes of managers as individuals. Yet, the feasibility of this preference is conditioned by the relocation of business. Since businesses are also moving, there is a convergence of interests that could contribute to a new model of urban flows, and therefore to a new pattern of urban settlements, under the conditions of efficient teleworking based on new ICTs. Nevertheless, it should be clear that the high-level decision-making centers and the main centers of innovation, be it in technology, research, or culture, seem to reproduce their traditional pattern of locating in a few centers supported by the best infrastructure and the symbolic value of their location.

Overall, the emerging spatial structure seems to be characterized by several simultaneous trends:

- Persistence of dominance of the largest metropolitan areas at the top of the hierarchy of wealth, power, knowledge, and culture. They are globally connected, and they continue to be magnets for attraction of capital and labor at the international level. They also are increasingly the most expensive areas to live and to do business.
- Intra-metropolitan decentralization of business, residence, and services, favored by the possibility of teleworking, and dense transportation systems.
- Some limited deconcentration of population and activities moving to medium-sized metropolitan areas and some rural areas.

Nonetheless, much of this movement is in fact a territorial expansion of large metropolitan areas that overflow into areas that can be connected by fast transportation and Internet networks. These areas may remain categorized as rural in the census, but they become part of the actual metropolitan areas. Furthermore, what characterizes the metropolitan region, in my analysis, is that it links up gradually with preexisting urban centers that become incorporated into the daily workings of the metropolis. Thus, what the growth of teleworking may be inducing over time is to increase the network of intra-metropolitan connections to form a much larger territorial expanse. The image of the "electronic cottage," working from a mountain top while cities fade into the past, continues to be a utopian vision, as an attempt to escape the unbearable reality of unsustainable megalopolises.

That is, unless emerging embryos of an alternative work and life culture pick up steam using teleworking as a tool of social transformation.

Digital nomads

"[Digital Nomads] are mobile professionals who perform their work remotely from anywhere in the world utilizing digital technologies" (Hannonen, 2020: 335). The concept was introduced by Makimoto and Manners in 1997. Although the phenomenon grew in importance throughout the years, it became highly significant after the Covid-19 pandemic, as companies released control over the workplace of their employees. In the US, the number of digital nomads grew by 49% from 2019 to 2020, reaching 10.9 million in 2020. Since that date it has continued to increase, reaching 15.5 million in the US and 35 million worldwide in 2023, creating a value of US$787 billion per year. This shows that its expansion is not limited to the effects of the pandemic. In fact, it is an indicator of a profound cultural change, in which freedom and discovery are paramount motivations for those highly skilled professionals whose knowledge and qualifications allow them to decide the time and space of their work. As Makimoto and Manners predicted, they are driven by "lifestyle freed by technology from constraints of geography and distance" (1997: 242). Lower cost of living than in the countries where they used to work is also a major consideration. In 2023, their average living costs in their chosen paradises were around US$1,700 monthly, while their average annual income was over US$120,000. Mobility and autonomy of their work does not lower their productivity. In fact, it increased by 4% over what it was in their previous location. Fifty-one percent of digital nomads come from the US, 8% from the UK, 5% from Russia, 5% from Canada, and 3% from Germany. Their preferences go to countries perceived as exotic, and definitely lower cost, with Mexico and Thailand topping the list, followed by Indonesia, Colombia, Costa Rica, Brazil, and Portugal. The top five most popular cities are Mexico City, Chiang Mai, Bali, Medellin, and Lisbon. Forty-six percent are self-employed, including freelancers, while 35% are employees working at distance. They are programmers, designers, creators, and work in technology, finance, cultural industries, and marketing. Two-thirds are single and are not interested in dating. Men in their thirties are the dominant group. Children and families are absent from their world. The pursuit of freedom, exemplified in their frequent traveling, is paramount in all dimensions of life. They rarely mix on a regular basis with the local population (Hermann & Paris, 2020; Clayton, 2021; MBO Partners, 2023). Many countries try to attract them because of their contribution to the economy and their technological know-how by offering special visas and fiscal exemptions. A case in point is Puerto Rico, which has succeeded in creating a large community of bitcoiners. Sarah Clayton has investigated in detail the fascinating experience of these bitcoiners in her doctoral dissertation (Clayton, 2021).

The phenomenon of digital nomadism is rooted in a culture of individual freedom that is achieved by superior knowledge and entrepreneurial spirit, essential value-holding assets in our society. Thus, it is bound to keep growing, with different manifestations, be it freelancing, remote work for companies, or self-programmed lives. It is statistically minor but culturally significant because it influences the imaginary of the young generation of techno-savvy professionals. Instead of being attracted by the amenities of the urban life of old cities, they decided to select for themselves the locations that were most desirable to them. So doing, they made them desirable and increased the outmigration of those professionals who had a choice to get away from the downtowns of the main metropolitan areas in developed countries. The spatial consequences of digital nomadism are complex to evaluate because trends and choices vary constantly, as the travel experiences become an endless search for new discoveries. However, we can already say that a small group of influential individuals has moved fully into a global space of flows, leaving behind a space of places where most people live, survive, and suffer in a degraded urban world from where the only ones able to escape are the new elites of the digital society.

Telework and cities in a global perspective

Most of the empirical analysis presented in this chapter depicts trends in the US, because much of the scholarly evidence that has been collected refers to this country. How has teleworking in the rest of the world impacted on the metropolitan structure during and after the Covid-19 pandemic? In Europe, both the size of the urban population and urban growth decreased during the pandemic, largely due to a decrease in net migration (Tricarico & De Vidovich, 2021). Larger cities experienced higher declines in population growth. Sixty-one of the 66 largest metropolitan areas experienced, on average, a drop of 93% in their growth rate. One in five cities had declines of over 1%, including 64 Spanish cities and 34 German cities. Sixty-three percent of European cities reported population loss in absolute terms during the pandemic; 28% were already losing population before the pandemic, 27% changed from growth to decline, with over half of Spanish, Italian, German, and Belgian cities being in this group. This included Berlin, Rome, and Madrid. Yet, overall, there has not been massive outmigration from metropolitan areas. In France, intra-metropolitan decentralization continued, with suburbs and exurbs growing in the periphery of the metropolitan areas.

In the UK, a decline in residential mobility has been observed, with greater mobility decreases in the most-dense areas (Rowe et al., 2023). Most city out-movers relocated in the nearby countryside, thus expanding the metropolitan area rather than reversing the migration flows. Indeed, when there is a nearby major city, outmigrants choose a new urban environment rather than the countryside. Only a small share of city out-movers changed location far from their city. For instance, just 4% moved from Inner London to other cities. In Italy, telework is very limited, and the decline in metropolitan population appears to be related to the reduction of international migration during the pandemic and to overall population decline (–6.7% in 2020, –4.3% in 2021). Pre-pandemic population increases reversed in northern and central Italy. In Spain, the territorial effects of the pandemic were more significant (González-Leonardo & Rowe, 2022). Thus, core cities increased outmigration and decreased immigration, while rural areas decreased outmigration (12.6%) and increased immigration by 20.5%. Madrid and Barcelona experienced substantial net migration declines (–12.4% and –14%), while low-density provinces had large gains in internal migration, particularly areas around Madrid. However, the study by Chapple et al. (2023) showed substantial downtown recovery in Europe by the Fall of 2022: economic activity increased in relationship to the pre-pandemic period by 175% in Barcelona, 150% in Madrid, 173% in Berlin, and 115% in Paris. The tourist industry seems to have played a role in the speed of this recovery. Therefore, metropolitan concentration and intra-metropolitan decentralization continue to be the dominant trends in the long term. Teleworking is largely intra-metropolitan.

In North America, the post-pandemic downtown recovery was more limited: 31% in San Francisco, 43% in Seattle, 46% in Montreal, as examples of a similar trend. The economy picked up in the main metropolitan areas after the pandemic. Yet, locations for residents and business changed their pattern, making plausible the hypothesis that remote work became more significant, and kept growing with a recovered economy.

Overall, spatial patterns in Europe after the pandemic do not seem to be different from those in the US, albeit with a greater strength of downtown areas in the post-pandemic period, perhaps due to their attractiveness for global flows of travelers.

In Japan, the pandemic slowed inter-prefectural migration toward Tokyo while remote areas gained population (Fielding & Ishikawa, 2021).

As for developing countries, which account for the large majority of world population, the few reliable studies (Gottlieb et al., 2021; Saltiel, 2020) evaluate at less than 10% the number of urban jobs that can be performed remotely. This considerably restricted the measures to prevent Covid-19 contagions in these countries. Remote work is limited to high-level managerial positions and clerical workers. Thus, the spatial patterns for most of the urban world continue to be determined by economic and social forces characterized by uneven territorial development. Most assets and job opportunities are concentrated in the largest metropolitan areas, which continue to attract massive flows of domestic and international migration that feed the expansion of gigantic metropolitan regions with extreme social inequality and environmental degradation. In 2023, the 20 largest metropolitan areas in the world are growing faster in population than the other metropolitan areas. The top ten areas house populations of over 20 or 30 million (Tokyo, Delhi, Jakarta, Shanghai, Manila, Seoul, Cairo, Kolkata, Mumbai, Sao Paulo) (World Population Review, 2023). In these areas, advanced telecommunication networks and Internet services connect their top managerial functions with similar locations around the world, forming a global network of communication to serve the strategically decisive nuclei of the global economy (Taylor & Derudder, 2016). The primary role of the global telecommunication infrastructure is to configure a space of flows for global capital and power, while removing the daily life of the dominant elites from the surrounding territories where most people live. There is, however, a special form of people's teleworking: the mobile phone networks of the informal economy that organize markets and provide jobs for the majority of the urban labor force. Therefore, the practice of teleworking is stratified like everything else between global flows and local places. However, local places are also connected by digital phone networks that provide platforms for survival in a sharply segregated metropolitan world.

References

Adrjan, P., Ciminelli, G., Judes, A., Koelle, M., Schwellnus, C., & Sinclair, T. M. (2022) "Working from home after Covid-19: Evidence from job postings in 20 countries," *SSRN Electronic Journal.* https://doi.org/10.2139/ssrn.4064191.

Aksoy, C.G., Barrero, J.M., Bloom, N., Davis, S.J., Dolls, M., & Zarate, P. (2022) "Working from home around the world," NBER Working Paper 30446. National Bureau of Economic Research. https://doi.org/10.3386/w30446.

Barrero, J.M., Bloom, N., & Davis, S.J. (2021) "Why working from home will stick," NBER Working Paper 28731. National Bureau of Economic Research. https://doi.org/10.3386/w28731.

Brynjolfsson, E., Horton, J.J., Ozimek, A., Rock, D., Sharma, G., & TuYe, H.-Y. (2020) "Covid-19 and remote work: An early look at US data," NBER Working Paper 27344. National Bureau of Economic Research. https://doi.org/10.3386/w27344.

Castells, M. (2010) "Globalisation, networking, urbanisation: Reflections on the spatial dynamics of the Information Age," *Urban Studies*, 47(13), pp. 2737–2745.

Castells, M., & Hall, P. (1994) *Technopoles of the World: The Making of 21st Century Industrial Complexes.* New York: Routledge.

Castells, M., Fernández-Ardèvol, M., Qiu, J.L., & Sey, A. (2006) *Mobile Communication and Society: A Global Perspective.* Cambridge, MA: MIT Press.

Chapple, K., Moore, H., Leong, M., Huang, D., Forouhar, A., Schmahmann, L., Wang, J., & Allen, J. (2023) "The death of downtown? Pandemic recovery trajectories across 62 North American cities," Research brief. School of Cities, University of Toronto; Institute of Governmental Studies, UC Berkeley.

Clayton, S. (2021) "Blockchain Migration: Narratives of Life Experiences in Puerto Rico at the Dawn of the Digital Era," Los Angeles, University of Southern California, PhD Dissertation in Communication.

Criscuolo, C., Gal, P., Leidecker, T., Losma, F., & Nicoletti, G. (2021) "The role of telework for productivity during and post-Covid-19: Results from an OECD survey among managers and workers," OECD Productivity Working Papers, Article 31. Available at: https://ideas.repec.org/p/oec/ecoaac/31-en.html.

Eurofound (2022) *The Rise in Telework: Impact on Working Conditions and Regulations.* Luxembourg: Publications Office of the European Union. Available at: https://www.eurofound.europa.eu/publications/report/2022/the-rise-in-telework-impact-on-working-conditions-and-regulations.

Fielding, T., & Ishikawa, Y. (2021) "Covid-19 and migration: A research note on the effects of Covid-19 on internal migration rates and patterns in Japan," *Population, Space and Place*, 27(6), e2499. https://doi.org/10.1002/psp.2499.

Florida, R., Rodríguez-Pose, A., & Storper, M. (2023) "Cities in a post-Covid world," *Urban Studies*, 60(8), pp. 1509–1531. https://doi.org/10.1177/00420980211018072.

Gareis, K., & Kordey, N. (2000) "The spread of telework in 2005." Available at: https://rauterberg.employee.id.tue.nl/presentations/Gareis%5B2000%5D.pdf.

González-Leonardo, M., & Rowe, F. (2022) "Visualizing internal and international migration in the Spanish provinces during the Covid-19 pandemic," *Regional Studies, Regional Science*, 9(1), pp. 600–602.

Gottlieb, C., Grobovšek, J., Poschke, M., & Saltiel, F. (2021) "Working from home in developing countries," *European Economic Review*, 133(C), 103679. https://doi.org/10.1016/j.euroecorev.2021.103679.

Hall, P. (1998) *Cities in Civilization*. London: Weidenfeld & Nicolson.

Hannonen, O. (2020) "In search of a digital nomad: Defining the phenomenon," *Information Technology & Tourism*, 22(3), pp. 335–353. https://doi.org/10.1007/s40558-020-00177-z.

Hansen, S., Lambert, P.J., Bloom, N., Davis, S.J., Sadun, R., & Taska, B. (2023) "Remote work across jobs, companies, and space," NBER Working Paper 31007. National Bureau of Economic Research. Available at: https://www.nber.org/system/files/working_papers/w31007/w31007.pdf.

Hermann, I., & Paris, C.M. (2020) "Digital nomadism: The nexus of remote working and travel mobility," *Information Technology & Tourism*, 22(3), pp. 329–334. https://doi.org/10.1007/s40558-020-00188-w.

Huws, U. (1991) "Telework: Projections," *Futures*, 23(1), pp. 19–31. https://doi.org/10.1016/0016-3287(91)90003-K.

Jarrahi, M.H., Lutz, C., Boyd, K., Oesterlund, C., & Willis, M. (2023) "Artificial intelligence in the work context," *Journal of the Association for Information Science and Technology*, 74(3), pp. 303–310. https://doi.org/10.1002/asi.24730.

Makimoto, T., & Manners, D. (1997) *Digital Nomad*. Chichester, UK: Wiley.

Mateyka, P.J., Rapino, M., & Landivar, L.C. (2012) "Home-based workers in the United States: 2010" (Current Population Reports P70-132; Household Economic Studies). US Department of Commerce Economics and Statistics Administration, US Census Bureau.

MBO Partners (2023) *Digital Nomads: Nomadism Enters the Mainstream*. MBO Partners. Available at: https://www.mbopartners.com/state-of-independence/digital-nomads/.

Myers, D. (2016) "Peak millennials: Three reinforcing cycles that amplify the rise and fall of urban concentration by millennials," *Housing Policy Debate*, 26(6), pp. 928–947. https://doi.org/10.1080/10511482.2016.1165722.

Ramani, A., & Bloom, N. (2021) "The donut effect of Covid-19 on cities," NBER Working Paper 28876. National Bureau of Economic Research. Available at: https://doi.org/10.3386/w28876.

Randazzo, M. (2022) "Digital Dating in the Networked Metropolis," Research paper, Seminar on the Network Society, Los Angeles, USC Annenberg School of Communication, unpublished.

Rowe, F., Calafiore, A., Arribas-Bel, D., Samardzhiev, K., & Fleischmann, M. (2023) "Urban exodus? Understanding human mobility in Britain during the Covid-19 pandemic using Meta-Facebook data," *Population, Space and Place*, 29(1), e2637.

Saltiel, F. (2020) "Who can work from home in developing countries?," *Covid Economics*, 7(2020), pp. 104–118.

Taylor, P.J., & Derudder, B. (2016) *World City Network: A Global Urban Analysis*, 2nd edn. New York: Routledge.

Tricarico, L., & De Vidovich, L. (2021) "Proximity and post-Covid-19 urban development: Reflections from Milan, Italy," *Journal of Urban Management*, 10(3), pp. 302–310. https://doi.org/10.1016/j.jum.2021.03.005.

World Population Review (2023) *World Population by Country 2023.* Available at: https://worldpopulationreview.com/ (accessed on 18 November 2023).

6 Human learning, computer learning, AI learning

The capacity to process information in order to learn from accumulated knowledge and experience is a fundamental feature of our species. This is the source of our ability to act upon our environment and onto ourselves to enact whatever our consciousness and will decide to do.

In modern societies, there has been an increasing formalization and institutionalization of the learning process in the education system. However, learning is much broader because it extends to all forms of learning, including learning in the family, at work, and in society at large. Formal education is also broader than learning because it includes the reproduction of norms and rules specific to the social context where education takes place.

My analysis in this book will focus on learning in the context of formal education, as the digitalization of the learning procedures is transforming the previous forms of teaching and learning, sometimes in tension with the established rules of educational institutions. Formal education offers a convenient vantage point of observation to examine the interaction between digital technologies of information and communication, and human learning. The key feature of formal education is that teaching mediates the learning process. And the fundamental transformation is the changing forms of interaction between teachers, machines, and students. In this regard, it may be useful to distinguish sequentially between different forms of this interaction.

Computers in education

The effects of digitalization on educational attainment have to be analyzed in a broader perspective by understanding the factors that favor academic performance, measured by scores in the classroom and at the country level. What has been established for a long time, starting with the classic study by Bourdieu and Passeron (1970), is that academic achievement is largely influenced by the social background of the students: the higher the educational level of the parents, the higher the success of their children. Educational level of parents correlates with income, and with the quality of the educational institutions attended by their children, the other elements that contribute to academic achievement. Yet, the educational level of families is in fact the most important factor, thus reproducing social stratification rather than correcting overall inequality.

Interpretations of this basic fact include the correspondence between the values of the teachers and the cultural level of the learned classes, as well as the ability of the family to provide help and support the students in their homework. A second major positive influence on students' performance is the quality of the teachers. Indeed, this is a consistent finding in the studies conducted in multiple countries and in diverse technological environments. The quality of the teachers is a direct function of the quality of their training in higher education institutions, of their salaries, and of their social prestige. Scandinavia, and particularly Finland, score high in these three factors, and so they have the highest quality primary education in the world (Castells & Himanen, 2002; Thrupp et al., 2023).

However, while these sociocultural factors continue to operate in the twenty-first century, the question remains open with regard to the potential changes introduced by digital technology in the education process. Time ago, there was a widespread belief that introducing computers in the classroom would help teaching and learning, potentially leveling the chances of students that were entering the digital world together, while their parents were less adapted to the digital generation. This belief led to policies such as "one laptop per child" as a lever of educational improvement, and ultimately, relative equalization of opportunities. It soon became clear that if computers were not connected to the Internet, they were little more than a typewriter attached to a calculator. Connecting the schools to the Internet thus became the real issue. This strategy met with the obstacle of lack of adequate connectivity infrastructure in many schools, particularly in low-income countries and neighborhoods.

Inequality in Internet access and in the quality of digital technologies continues to be a major issue, as I will show in Chapter 7. However, once schools were updated to incorporate laptops with Internet connection, scholarly research discovered that there is no magic formula to improve education by simply introducing digital technology. Indeed, in surveys both in the US and at the international level, the large majority of teachers agree that the use of ICTs was a priority to improve their work. However, in a 2018 International Computer Literacy Survey, almost half of teachers complained that they did not have enough technical support, and only 41% considered that there was not enough time to prepare the lessons with the new methodology. Moreover, in spite of the favorable attitude of teachers toward digital technology, the same survey in a number of countries found that the average percentage of teachers using ICTs at the school was 48%, reaching 50% in the US, and with some countries ranking much lower, such as Germany (23%), Italy (35%), or Chile (25%) (National Center for Education Statistics, 2018). In fact, only a minority of people surveyed in a number of countries considered that their formal education had given them the technology knowledge they need (39% for the US, 31% for the UK, 25% for Germany), with China being the exception (68% feel they acquired the necessary skills and knowledge) (Statista, 2018). This may be the outcome of governments accounting for only 11% of funding for educational technology innovations, in contrast to the private sector (40%) or even NGOs (46%) (Brookings Institution, 2019).

Research on the relationship between use of computers at school and academic achievement counters the optimism about the positive role of digital technology. The main reliable source in the measurement of academic achievement, the OECD PISA studies that compare a number of countries, found no positive correlation between use of computers in schools and educational outcomes. Moreover, in most countries, the use of computers in the classroom was negatively correlated with academic performance. On the other hand, in some countries, use of computers at home proved beneficial for math and science learning. A possible hypothesis is that the school setting was not favorable to a synergistic contribution between teachers and students. Indeed, students who do not use their digital devices in their reading classrooms have a better academic performance than those who use them, in every region of the world, with the important exception of North America. In North America, Sweden, and Australia, using digital technology for more than 60 hours per week is positively associated with higher scores. Thus, there is clearly a wide diversity of technology effects that appear to be associated with the level of technological development. However, in most of the world, low use of technology correlates positively with achievement (OECD, 2018). Another OCED study asserts that "As in earlier studies, the frequency of technology use in schools correlates negatively with math, reading, and science achievement in the large majority of countries," a finding that is supported and analyzed by a number of academic articles (Shewbridge et al., 2005; Woessmann & Fuchs, 2007; OECD, 2015).

On the other hand, an older study concluded that in most countries the use of computers *at home* correlates positively with science scores (OECD, 2010). How can we make sense of the discrepancy of findings, particularly since the most used data come from the same source (OECD, 2018) at different points in time? I would suggest that educational institutions have met difficulties in deepening the incorporation of digital technologies successfully. In the research program I directed on a representative sample of secondary schools in Catalonia, we reached a meaningful conclusion: in spite of the fact that the schools were connected to the Internet, teachers and students would perform better when working at home on their own, while their work with Internet at the school was not frequent and constrained by institutional rules (Momino et al., 2008). Yet, working at home using the Internet is not advised by educators out of fear that unsupervised students may be distracted by playing video games, particularly boys, who spend more time than girls playing rather than studying (OECD, 2018; Statista, 2019). Excessive video-gaming negatively affects academic performance (OECD, 2018: 35). However, contrary to the usual perception of adults, playing video games seems to improve performance in digital reading, and overall digital skills when limited to a moderate amount of time.

The fact remains that the majority of findings in the stream of research on computers and academic advancement in school settings point to disappointing results of the contribution of technology per se to better educational outcomes. Thus, Karlsson (2022) found a negative association between computer use and test scores in primary schools. This negative impact of technology on academic achievement was larger among low-performing pupils. What about computers for academic work at home? In an even more shocking finding, still using international survey data from OECD PISA, Agasisti et al. found that "the use of ICT at home for school-related tasks results in the average student achieving lower scores and that this negative association is more pronounced for low achievers and even more so for high achievers" (2020: 601–620). Belgium and the Netherlands were exceptions, showing positive effects. Yet, all other countries in the survey showed negative associations, particularly Germany.

How can we make sense of the recurrent finding showing no relationship or even negative relationships between use of the Internet and learning in school-related tasks? An interesting hypothesis, elaborated by OECD experts, suggests that "In the past, students could find clear and often singular answers to their questions in carefully curated and government-approved textbooks, and they could generally trust those answers to be true" (my emphasis).

Today, they will find thousands of answers to their questions online, and it is up to them to figure out what is true and what is false, what is right and what is wrong. While in many off-line situations readers can assume that the author of the text they are reading is competent, well informed, and benevolent, when reading online blogs, forums, or news sites, readers must constantly assess the quality and reliability of the information, based on implicit or explicit cues related to the content, format, or source of the text. This is not exactly a new phenomenon, but the speed, volume, and reach of information flows in the current digital ecosystem have created the perfect conditions for fake news to thrive, affecting public opinion and political choices. In this "post-truth" climate, quantity seems to be valued more than quality when it comes to information. Assertions that "feel right" but have no basis in fact become accepted as truth. Algorithms that sort people into groups of like-minded individuals create social media echo chambers that amplify views, and leave individuals uninformed of and insulated from opposing arguments that may alter their beliefs. "There is a scarcity of attention, but an abundance of information" (OECD, 2018). The same PISA study found that less than 10% of students in OECD countries were able to differentiate between fact and opinion.

In line with these arguments a striking new development in educational policy took place in June 2023. Sweden, one of the first countries to introduce computers in public schools, reversed the practice, and downplayed the role of the Internet in learning while restoring the widespread use of textbooks.

Lost in the fog of the Internet galaxy, and with little training in critical reading, students become largely dependent on the guidance of their teachers. At the same time the worsening of the conditions under which educational institutions perform creates a major obstacle for teachers to fulfill their traditional role, while the digitally savvy generation lives in the fantasy of being empowered by their unlimited access to information without being able to evaluate the accuracy. It follows a crisis of the learning process increasingly dominated by disinformed bewilderment.

Distance learning

Distance learning is a traditional form of education, particularly in higher education and in professional training. Yet, it has been transformed by the use of the Internet. The first major university to teach 100% over the Internet was the Open University of Catalonia (UOC), which started operations in September 1995, the same year the commercialization of the World Wide Web began. Currently the UOC offers a broad variety of programs and degrees to over 60,000 students. Some of the best distance education institutions, such as the Open University in the UK and the Open University in the Netherlands, gradually shifted to 100% Internet teaching. In the twenty-first century, there was a massive expansion of distance learning, both public and private, at all levels of education, but mainly in higher education and employee training. The total number of enrolments reached 47 million in 2017. Naturally, the trend increased substantially during the Covid-19 pandemic, reaching 189 million in 2021 (WEF, 2022). In the US, the percentage of undergraduate students at degree-granting postsecondary institutions who enrolled exclusively in distance education courses reached 52% of the total. In the case of private for-profit universities, it was much higher: 89% of the total (National Center for Education Statistics, 2021). It appears that for-profit universities believe that Internet-based education offers the possibility to teach to large numbers while saving costs by automating the learning process instead of hiring teachers. However, cost cutting depends on the quality of the service provided. Internet-based teaching that uses personalized interaction and continuous evaluation of the learning process can in fact be more expensive, because it requires a vast network of highly qualified educators. Yet, the large majority of online courses offer substandard quality. In most Latin American countries, the paradox is that public in-presence universities, usually of better quality than most private institutions, enroll middle-class students paying low tuition because of their better performance in academic tests, while lower-income groups tend to only access more expensive for-profit online universities. The debate concerning the quality of online education rages all over the world, and online education is usually considered to be inferior to traditional education by the academic establishment.

Cutting across the fog of a polarized debate dominated by corporatist ideology and business interests, I can propose three main conclusions. First, distance education seems to be better suited to professionals who are already in the labor market and need to update or acquire a specific skill or receive a new degree as they search for a promotion.

Secondly, for standard undergraduate education, and for advanced research degrees, in-presence education is preferable, particularly because of the role played by the social environment among young students.

Thirdly, in those cases where the location of students makes it difficult to travel to a school, and even more so when there is an emergency, such as a pandemic (probably a recurrent event in the coming years), there is no option other than distance education. If so, it is essential to improve the quality, the methodology, and the connectivity infrastructure for a fruitful deployment of education and vocational training on a large scale.

Furthermore, the existence of distance education may increase educational inequality if low-income students do not have access to quality infrastructure, or any infrastructure at all. This is particularly dramatic in the case of primary and secondary education. Thirty-one percent of students in these educational levels worldwide cannot be connected digitally, either because of lack of public infrastructure or lack of family resources. This is a particularly acute problem in Africa, South Asia, and Latin America (Brookings Institution, 2020). According to UNICEF, at least 463 million students cannot access distance learning (UNICEF, 2020).

Artificial intelligence, machine learning, ChatGPT

The process of learning and, more specifically, the formal education system are being deeply impacted, according to most experts in the field, by the applications of artificial intelligence (Holmes & Tuomi, 2022), understood as the programming of algorithms to refine existing knowledge and imitate forms of reasoning and decision making used by humans. In recent years these algorithms have shown a giant leap in performance, expressed for instance by Google's AlphaGo beating the human world Go champion in October 2021. However, the apparent simplicity of this conception of AI is contested by Kate Crawford, one of the leading analysts of AI, who argues in a powerful book that "AI is neither artificial nor intelligent" (Crawford, 2021: 7). Because it is a human production, generated and applied in specific social contexts, "each way of defining artificial intelligence is doing work, setting a frame for how it will be understood, measured, valued, and governed" (Crawford, 2021: 7). Accepting this perspective, I will be focusing here on the implications of specific algorithms deployed in the process of learning, which ultimately amounts to asserting the potential of machine learning on human learning, that is, understanding by machine learning the use of algorithms that enable computers to make predictions or decisions based on a large amount of data without the need for explicit programming.

Machine learning operates in two different modalities: supervised and unsupervised learning. In the first case, humans guide the computer through a process of pattern recognition and decision making. In the case of unsupervised learning, computers are provided with programs able to identify patterns within large collections of data through which they develop rules toward interpreting future data without direct human intervention. I emphasize that the critical foundation of this technology is the datasets that provide the raw material with which the algorithm can be trained. And these data are not neutral; they are collected and organized according to interests, values, and purposes that the machine cannot evaluate unless instructed to do so.

At the forefront of the discussion on AI in education at the time of writing is the use of LLMs that are powered by massive databases of text from which they manage to extract very intricate patterns, such as the rules underlying human language. Perhaps the most popular of these models, as I mentioned in the Introduction, is ChatGPT (short for Generative Pre-Trained Transformer). Although the operating principle is the same, their training may induce substantial differences. For instance, in the case of a ChatGPT competitor, Claude GPT (designed by Anthropic), the company claims that a distinctive characteristic that might give it a competitive edge is that it is "much less likely to produce harmful outputs" (Edwards, 2023). This is achieved through the process of cleaning datasets

of harmful content and through training, where the values of programmers can constrain the way in which the machine will create outputs.

The term Generative refers to the capacity of the machine to generate new
patterns and new arguments that were not included in the instructions of
the program or present within the training datasets that it was exposed
to. It is this feature that triggers all kinds of human hallucinations about
the potential autonomy of the machines, and thus the social alarm that
accompanies the rise of what has been labeled as the "algorithmic culture."
In this chapter I will focus on the issues surrounding AI in learning, as the
broader discussion on the ethical and political implications of artificial
intelligence is treated elsewhere in this volume. Before considering the
specific impact of ChatGPT, it is worth noting the wide uses of AI in
education in programs such as adaptive learning, smart campus, teacher
evaluation, intelligent tutoring assistants, and virtual classrooms. Much of
the reliable research concludes that "AI has a beneficial effect on both the
quality of instruction provided by teachers and on learning outcomes of
students" (Alam, 2022: 395–406), although the jury is still out as this field
of study evolves at a fast pace.

As for ChatGPT, examining the limited evidence on the uses of ChatGPT
in education available at the time of writing, it appears that it offers a wide
range of possibilities: it can provide a platform of information on available
data and analyses on a given topic, or it can construct one or several arguments depending on the purpose of the request. In this case it is a useful
tool for teaching, learning, and research. The guidance of teachers in all
cases seems to be essential.

This may explain why, at least in the US in 2023, 22% of students and
40% of teachers reported to use it regularly. And the vast majority of
them think that schools should bring themselves up to date in the use
of new technologies (Walton Family Foundation, 2023). On the other
hand, a majority of teachers complain that they have little guidance from
the school, 43% consider that these tools make their jobs more difficult,
and half of them have experienced the problem of students cheating in
their assignments (USA Today, 2023). Of adult students, 89% have used
ChatGPT to do their homework, including 48% who reported using it
to perform a test. Others admitted to using it to write essays or to create
a paper outline. However, in multiple instances, they use ChatGPT as an

assistant, for instance for translation, which helps advance their academic skills (Zhang et al., 2023).

In spite of the diversity of uses, some of them highly beneficial for the students, secondary schools in the US have reacted very negatively, emphasizing the risk of cheating as the main implication of ChatGPT. Key school districts like New York, Los Angeles, Seattle, and Virginia have banned its use (Johnson, 2023).

In the case of higher education, there is an array of policies: some aim at completely banning GPTs; others allow their use for certain tasks; others require permission and citing or acknowledging ChatGPT as a source of information (Caulfield, 2023). Overall, institutional distrust of students seems to be the most prevalent reaction to technological change. This shows the conditioning of AI uses by the values inherent to the context in which it is used. Most educational institutions are very conservative in their pedagogy, and emphasize control and authority over the students, ultimately implemented in the sacred nature of exams. In many cases these exams are based on repeating memorized information. In an age in which all information is accessible on the Internet this seems outdated, and so the most important educational outcome is to develop the ability of students to think critically and to recombine information in the production and expression of their own understanding.

This gap between the digital culture and traditional values creates a tension in some school systems because it means more work for the teachers and less authority for the school. There is an obvious need to experiment with a new pedagogy adapted to the culture and technology of our time. To be sure, within the current pedagogical structures, the use of ChatGPT has the risk of cheating in assignments, which is a problem because it is unfair for other students and could lead to a lack of incentives for learning.

It is worth noting that AI-powered systems also provide tools to detect the origin of a text. Detectors of plagiarism have been in place for a long time and disciplinary actions to sanction plagiarism are widely enforced. It may be the same case in the near future with GPT programs. GPTZero and Originality.AI are among the most used detector programs in 2023. With the safeguard of AI control of AI uses, educational institutions could allow teachers and students to experiment freely with AI assistance while keeping their autonomy to take their thinking above and beyond tasks that have or can be automated. This is a frequent practice in a number of private companies with their employees. Some studies already point toward the academic advantages of using the new machine learning contributions by making them complementary and not substitutive of human learning (Dwivedi et al., 2023).

The virtual classroom

Looming on the horizon of educational transformation, the "virtual classroom" merges physical and virtual worlds, by using virtual reality and augmented reality, and combining in-presence communication and Internet-based interaction. Students benefit from graphic displays of simulation models in science, or from guided visits to distant museums, or from experimentation of their own ideas in interaction with GPT models.

Assisted learning and tutoring robots mediate between teachers and students, and among student groups, while exploring the hypertext of images and text that display the stock of human knowledge. In this context, and under proper supervision, entertainment video games could be a positive factor in the learning process (Martinez et al., 2022).

These virtual classrooms do exist in a few elite institutions. It would seem out of touch with the real world to build these new techno-pedagogical systems in a world in which at least half a billion children cannot connect to any external source of information and most of the schools barely survive with the indispensable help of textbooks. Yet, these images of a new learning environment are not just a futuristic fantasy, but a blueprint of what human learning could be if people, properly assisted by machines, really mattered.

References

Agasisti, T., Gil-Izquierdo, M., & Han, S.W. (2020) "ICT use at home for school-related tasks: What is the effect on a student's achievement? Empirical evidence from OECD PISA data," *Education Economics*, 28(6), pp. 601–620. https://doi.org/10.1080/09645292.2020.1822787.

Alam, A. (2022) "Employing adaptive learning and intelligent tutoring robots for virtual classrooms and smart campuses: Reforming education in the age of artificial intelligence," in R.N. Shaw, S. Das, V. Piuri, & M. Bianchini (eds), *Advanced Computing and Intelligent Technologies*. Singapore: Springer Nature, pp. 395–406.

Bourdieu, P., & Passeron, J.-Cl. (1970) *La Reproduction. Éléments d'une théorie du système d'enseignement*. Paris: Les Editions de Minuit.

Brookings Institution (2019) "How ed-tech can help leapfrog progress in education," *Brookings*, 20 November. Available at: https://www.brookings.edu/research/how-ed-tech-can-help-leapfrog-progress-in-education/ (accessed on 17 May 2023).

Brookings Institution (2020) "Unequally disconnected: Access to online learning in the US," *Brookings*, 22 June. Available at: https://www.brookings.edu/blog/education-plus-development/2020/06/22/unequally-disconnected-access-to-online-learning-in-the-us/ (accessed on 29 May 2023).

Castells, M., & Himanen, P. (2002) *The Information Society and the Welfare State: The Finnish Model*. Oxford: Oxford University Press.

Caulfield, J. (2023) "University policies on AI writing tools|Overview & list," *Scribbr*. Available at: https://www.scribbr.com/ai-tools/chatgpt-university-policies/ (accessed on 16 May 2023).

Crawford, K. (2021) *The Atlas of AI: Power, Politics, and the Planetary Costs of Artificial Intelligence*. New Haven, CT: Yale University Press.

Dwivedi, Y.K. et al. (2023) "'So what if ChatGPT wrote it?' Multidisciplinary perspectives on opportunities, challenges and implications of generative conversational AI for research, practice and policy," *International Journal of Information Management*, 71, 102642. https://doi.org/10.1016/j.ijinfomgt.2023.102642.

Edwards, B. (2023) "Anthropic introduces Claude, a 'more steerable' AI competitor to ChatGPT," *Ars Technica*. Available at: https://arstechnica.com/information-technology/2023/03/anthropic-introduces-claude-a-more-steerable-ai-competitor-to-chatgpt/ (accessed on 30 May 2023).

Holmes, W., & Tuomi, I. (2022) "State of the art and practice in AI in education," *European Journal of Education*, 57(4), pp. 542–570. https://doi.org/10.1111/ejed.12533.

Johnson, A. (2023) "ChatGPT in schools: Here's where it's banned—and how it could potentially help students," *Forbes*, 31 January. Available at: https://www.forbes.com/sites/ariannajohnson/2023/01/18/chatgpt-in-schools-heres-where-its-banned-and-how-it-could-potentially-help-students/ (accessed on 30 May 2023).

Karlsson, L. (2022) "Computers in education: The association between computer use and test scores in primary school," *Education Inquiry*, 13(1), pp. 56–85. https://doi.org/10.1080/20004508.2020.1831288.

Martinez, L., Gimenes, M., & Lambert, E. (2022) "Entertainment video games for academic learning: A systematic review," *Journal of Educational Computing Research*, 60(5), pp. 1083–1109. https://doi.org/10.1177/07356331211053848.

Momino, J.M., Sigales, C., and Meneses, J. (2008) *La Escuela en la Sociedad Red*. Barcelona: Ariel.

National Center for Education Statistics (2018) "Teaching with technology: US teachers' perceptions and use of digital technology in an international context," *NCES Blog*. Available at: https://nces.ed.gov/blogs/nces/post/teaching-with -technology-u-s-teachers-perceptions-and-use-of-digital-technology-in-an -international-context (accessed on 30 May 2023).

National Center for Education Statistics (2021) "The NCES Fast Facts Tool provides quick answers to many education questions," *National Center for Education Statistics*. Available at: https://nces.ed.gov/fastfacts/display.asp?id= 80 (accessed on 24 October 2023).

OECD (2010) *Does Computer Use Increase Educational Achievements? Student-Level Evidence from PISA*. Paris: OECD. Available at: https://www .oecd-ilibrary.org/economics/does-computer-use-increase-educational -achievements-student-level-evidence-from-pisa_eco_studies-2010 -5km33scwlvkf (accessed on 29 May 2023).

OECD (2015) *Students, Computers and Learning: Making the Connection*. Paris: OECD. https://doi.org/10.1787/9789264239555-en.

OECD (2018) *PISA 2018 Insights and Interpretations*. Paris: OECD. Available at: https://www.oecd.org/pisa/PISA%202018%20Insights%20and%20Interpretations %20FINAL%20PDF.pdf (accessed on 16 May 2023).

Shewbridge C., Ikeda M., & Schleicher A. (2005) *Are Students Ready for a Technology-Rich World? What PISA Studies Tell Us*. Paris: OECD. Available at: http://www.oecd.org.libproxy2.usc.edu/education/school/programmeforinter nationalstudentassessmentpisa/35995145.pdf.

Sigales, C. et al. (2007) *La escuela en la sociedad de la informacion*. Barcelona: Ariel.

Statista (2018) "Infographic: Education struggling to keep up with digital advances," *Statista Infographics*. Available at: https://www.statista.com/chart/ 13017/education-struggling-to-keep-up-with-digital-advances (accessed on 30 May 2023).

Statista (2019) "Daily time spent gaming among kids by gender 2019," *Statista*. Available at: https://www.statista.com/statistics/1128307/video-gaming-kids -gender/ (accessed on 29 May 2023).

Thrupp, M., Seppänen, P., Kauko, J., & Kosunen, S. (eds) (2023) *Finland's Famous Education System*. New York: Springer.

UNICEF (2020) "Infographic: At least 463 million students cut off from remote learning," *Statista Infographics*. Available at: https://www.statista.com/chart/ 22799/number-of-children-with-and-without-access-to-remote-learning -programs (accessed on 29 May 2023).

USA Today (2023) "ChatGPT in the classroom: Here's what teachers and students are saying," *USA Today*, 1 March. Available at: https://www.usatoday.com/ story/news/education/2023/03/01/what-teachers-students-saying-ai-chatgpt -use-classrooms/11340040002/ (accessed on 17 May 2023).

Walton Family Foundation (2023) "Teachers and students embrace ChatGPT for education," *Walton Family Foundation*. Available at: https://www.wa ltonfamilyfoundation.org/learning/teachers-and-students-embrace-chatgpt -for-education (accessed on 29 May 2023).

WEF (2022) "These 3 charts show the global growth in online learning," *World Economic Forum*. Available at: https://www.weforum.org/agenda/2022/01/online-learning-courses-reskill-skills-gap/ (accessed on 17 May 2023).

Woessmann, L., & Fuchs, T. (2007) "Computers and student learning: Bivariate and multivariate evidence on the availability and use of computers at home and at school," CESifo Working Paper Series 1321. Retrieved from http://papers.ssrn.com.libproxy2.usc.edu/sol3/papers.cfm?abstract_id=619101.

Zhang, C. et al. (2023) "One small step for generative AI, one giant leap for AGI: A complete survey on ChatGPT in AIGC era," *arXiv*. Available at: http://arxiv.org/abs/2304.06488 (accessed on 17 May 2023).

7 Digital divides: territory, gender, age, class, ethnicity, cultures

The digitalization of society has spread globally at a fast pace, and it is projected to expand further in the short term (2030), as I documented in the Introduction to this volume. However, the diffusion of technological affordances and users' access and capabilities vary in step with the sources of inequality around the world. Moreover, the United Nations Agenda 2030 redefined its goals for digitalization from aiming for "universal access" to "universal and *meaningful* access," a distinction that we aim to make more precise when examining the empirical record in this chapter.[1]

Territorial divides

In the age of globalization and Internet networking, digital inequality between countries remains, largely associated with the wealth of the country. The percentage of Internet users over the global population in 2022 was 66%, but it reached 92% in the high-income countries, 79% in the upper-middle-income countries, falling to 56% for the lower-middle-income group, and to 26% for the low-income countries. There is also a difference of penetration for smartphones, although there is a high proportion of users of smartphones everywhere, the highest percentage being in Europe, with 92%, the same percentage as in the US. Mobile cellular subscriptions reach 86% in Africa and over 100% in most regions of the world, with 121% the figure in Europe, illustrating the unequal distribution, whereby some people have several subscriptions while others, even if they are a small minority, have none. Territorial

---

[1]   For simplicity, references for statistics on ICT use will be omitted in the text. They are included in the references section, and are typically from ITU when it comes to aggregates. Exceptionally, data is sourced from Statista or the Pew Research Center.

inequality extends, in every country, to the sharp disparity between urban and rural areas, while the need for broadband connectivity is more acute for rural localities. Indeed, a study by Galperin et al. (2022) on rural Ecuador in 2011–19 illustrates this. The researchers found substantial gains in income, and a modest improvement in employment, associated with the introduction of high-speed broadband that facilitated access to critical market information, and helped to diversify employment. In a larger context, a statistical analysis of 24 Latin American countries in 2020 calculated that an increase of 1% in fixed broadband penetration yields an increase of 0.08% of GDP, and a 1% increase in mobile connectivity increases GDP by 0.15% (Ziegler et al., 2020). And yet, in Latin America, where rural poverty is double that of urban poverty, Internet penetration in rural areas stood at 37% in 2020 in contrast to 71% in urban areas. However, this is somewhat compensated by the penetration of smartphones in rural areas, which reached 71% in 2020 (Ziegler et al., 2020). Worldwide, while urban areas displayed an 82% Internet penetration in 2022, rural areas lagged behind at 46%. The rural–urban cleavage persists everywhere, but is less pronounced in highly developed countries: in the US in 2021, Internet use in urban areas was 95%, and in rural areas 90% (Statista, 2021). In the UK in 2020, Internet penetration in urban areas was 98% and in rural areas 94% (Statista, 2020a; 2020b), although here again mobile phone penetration at 130% ensures communication in most of the country. China presents an intermediate situation, with 77% urban penetration and 46% rural access to the Internet in 2020 (ESGN, 2021). In all cases the gap in quality of the connection is substantial, as the rural infrastructure lags behind the networks deployed in the main urban centers.

There is a second territorial divide, namely the location of the infrastructure for digital communication, tantamount to a differential access to the "cloud," because the cloud materializes in servers that store and direct data traffic originated anywhere. As of 2018, the large majority of international traffic from Latin American countries went to servers in North America (73%) and Europe (10%). In the case of African countries, 56% of the international traffic went to North American servers and 32% to Europe. Even Europe is not self-sufficient in its servers, as 43% of traffic used North American servers, while 56% of international traffic went to another European country. China relies on its own servers for domestic traffic, but its global access may also be dependent on North American servers that manage 49% of international traffic from Asia–Pacific, which includes Chinese data. The hubs of international traffic are located in a few areas to serve entire regions of the world: four of the five highest capacity international routes in Latin America are connected in Miami (Telegeography, 2018). The disparities in infrastructure location and control are accentuated by the ownership of exclusive undersea cables by major digital companies such as Google, Facebook, Amazon, and Microsoft that account for the largest share of the increase in global bandwidth in recent years (Satariano, 2019). Preferential location to the hubs of the global infrastructure has substantial consequences for the design of national technology policies, and obvious implications for national security, thus decreasing sovereignty in the majority of countries as we transition to a more digital environment (Ortiz Freuler, 2020).

The deployment of advanced telecommunications infrastructure is also differential between countries, as illustrated by the global distribution of 5G networks (Statista, 2023). With the significant exception of sub-Saharan Africa, by 2020 most countries had launched, deployed, or invested in 5G. For instance, most of South America had already launched the networks and the remaining countries had deployed the technology. China is among the most developed users of 5G because Huawei was critical in the design of the technology, and set up 70% of the base stations in the world as of 2023 (Parzyan, 2023). But Indonesia also is in the leading group while Russia is in an earlier phase of development. The geography of telecom networks is a major indicator of the extent of inequality in the process of digitalization worldwide.

The gender gap

The gender gap in the digital world goes beyond the difference between men and women in Internet access. Indeed, in 2022, the gap was limited in terms of global statistics: 70% of men were daily users, a somewhat higher proportion than 63% of women. But this difference masks substantial variation among countries, depending on the geography of patriarchalism. In the UK in 2020, men were more likely to use the Internet daily (90% versus 89%; ONS, 2021), while the figures reported for the US in 2023 were similar (94% for men and 93% for women). In Spain in 2021, the percentages of users were close (77% for men, 75% for women) (Garín-Muñoz et al., 2022). But the differences are greater in certain regions of the world. The proportions of Internet use for men and for women were respectively 83% and 83% for the Americas, 89% and 90% in Europe. However, the gap was larger in other regions, with 67% and 61% in Asia–Pacific, 75% and 65% in the Arab States, and 46% and 34% in Africa, respectively.

Furthermore, the world's Internet use among women in high-income countries was 92%, in sharp contrast with the low-income countries where only 21% of women use the Internet. Even in China, now a rather developed country widely using mobile phones, in 2020 74% of men owned smartphones but only 54% of women did. In sum, the level of development of countries specifies the digital divide: the more developed the country is, the greater the trend toward shrinking the gender gap. And vice versa. This is clearly related to the condition of women in most of the Global South, widely submitted to patriarchal domination.

However, access is not the only dimension of the gender gap in the digital age. Pay in the digital industries is substantially lower for women, controlling by the type of occupation: it is 20% lower than men in the US, and 33% lower in less developed countries (ILO, 2019). The leading companies in the digital industry employ a minority of women: only 17% of Facebook's employees are women, 19% for Google, 23% for Apple (Chakravorti, 2017). This is largely the result of the low proportion of women graduating in science, math, engineering, and technology, about 25% of all graduates in the US. One of the leading researchers in this field, Cecilia Castano has thoroughly examined in a series of studies the roots of this low presence of women in technology occupations, and found it in the triage operated in high schools toward learning for gendered labor markets, and in the discriminatory culture prevalent in a number of engineering schools (Castano, 2005a, 2005b, 2015, 2021; Castano et al., 2010; Castano & Webster, 2011, 2014).

Furthermore, women attempting entrepreneurship in the digital industry face considerable difficulties. According to OECD (2018) studies, women receive 23% less funding than their male counterparts to start a company. There are more subtle, and more damaging, differences in gender access to social media. In general, around the world, more women than men use social media. An internal study commissioned by Instagram found that social media made teens feel worse about their bodies and that they blame the platform for anxiety, depression, and suicidal thoughts (Wetsman,

2021). Some studies in the UK have shown that young girls (11–13) are facing greater emotional challenges than their male classmates (Fink et al., 2015). Large-scale surveys in the US report teen girls being subjected to online bullying and aggression more often than their male counterparts (Pew Research, 2022). Meanwhile, a growing number of studies support the association between loneliness, depression, and suicidal thoughts with the use of smartphones and social media (Shensa et al., 2018; Allcott et al., 2020; Twenge, 2023). Furthermore, the deficit in the presence of women in digital design (e.g., video games) tilts the content toward a predominantly male, and often violent, culture that appeals to young high-testosterone males and excludes other habits and visions of the world produced by women's specific experience. In my conversation in 2019 in Shenzhen with Mr. Ma, founder and CEO of Tencent, the Chinese company with the largest market share of video games in the world, he expressed openly his concern for this gender bias, and his desire to gradually change the content of the games.

In sum, the digital gender gap and its unwanted consequences result from the differential levels of development of societies, the systemic discrimination against females, their cultural apartheid in certain roles and occupations, and the still sharply contrasted geography of patriarchalism, with its multiple sexist corollaries.

The grey divide

Differential use, access, and quality of Internet connection depending on age is the most significant divide in this early digital society, because it is likely that as the people in the age cohorts over 65 and 75 years, who grew up before the advent of the Internet, pass away, the overall digital divide in terms of access will dwindle. Indeed, the proportion of the world population over 60 years old in 2021 was 12%, while the proportion of persons older than 65 in the population of Internet users was 6% (Statista, 2021). On the other hand, the proportion of the population under 20 years is 33%, while the percentage of Internet users in the group 25–34 is about one-third of all Internet users, and in the group 18–24 was 18%. However, this demographic trend does not imply that the digital divide will disappear by biological law, because in 2022, while the percent of Internet users in the age group 15–24 reached almost 100% in the high-income countries, in the low-income countries it was still under 40%. The wealth of countries and the uneven distribution of technological capacity will still translate into a digital divide for the years to come. The situation is increasingly different in developed countries, yet a grey divide still exists. Thus, in the US, in 2021 in the group 18–29 years old, 99% are Internet users, while among 65 years and older, the proportion is 75%.

Indeed, a significant number of seniors have not joined the digital age: in the UK, 26% of people 65–74 and 61% of those older than 75, do not regularly use the Internet. In the US, 25% of people older than 65 are not Internet users. In the European Union, the use of the Internet, in spite of its availability, varies considerably for the age group 65–74, from 90% in Denmark and 70% in Spain, to 40% in Portugal and 25% in Croatia. Furthermore, the stream of studies conducted over the years by Mireia Fernández-Ardèvol and her international network of researchers have questioned the linearity of the relationship between age, use, and meaning of mobile Internet for the older groups of the population (Fernández-Ardèvol et al., 2017; 2023; Fernández-Ardèvol, 2020; Beneito-Montagut et al., 2022). They have shown the key role of smartphones in the communicative practices of the elderly, mainly because of the multimodality of the devices. On the basis of qualitative studies, they emphasized what they call techno-diversities, in terms of the use of communication devices over the life cycle, particularly the significant change of use and meaning after 75 years, what they call the "fourth age" group. For many elderly, the nonuse or the selective use of Internet and smartphones is not necessarily due to lack of digital capabilities, but to a mechanism of empowerment, by deciding what, with whom, and how to communicate. In other words, the inclusion/exclusion in the networks of digital communication is not the result of unilateral policies of companies (such as the pricing out of low-income elderlies) but of an interaction between life histories, and adaptation to a disadvantaged position, worsened by hyper digitization, ultimately resulting in the affirmation of the right to decide their own life, in spite of physical and cultural disabilities. The right to be left alone, although not excluded from society, is becoming the new frontier of digital rights, beyond the strategies and interests to expand markets for telecommunication business.

Digital divide by class

Class, as usually defined in stratification analysis, refers to the differentiation of people in terms of their wealth (income, assets), level of education, and occupational status. In fact, these three categories correlate. In all countries, there is a direct relationship between the high ranking in each one of these categories, and access to and use of the Internet (Mubarak et al., 2020). Moreover, even in contexts where social hierarchies are defined in cultural and institutional terms, as in the case of caste in Indian society, income and education remain the key factors in explaining the caste stratification in Internet use (Rajam et al., 2021).

The educational background of the family reproduces the social differentiation of their children in their Internet access, both at home and at school. In the US, for instance, in 2021, 98% of children in the group 13–18 years old with parents with college education or higher have home Internet access, while among those children with parents with an educational level of high school, 79% have home Internet access (NCES, 2021). In low-income countries, the difference in access according to family background is much larger.

However, as smartphone ownership spreads around the world, it is necessary to update the understanding of digital inequality. Indeed, in 2023, 86% of people in the world had a smartphone, and 91% had a mobile phone. Considering that 94% of global Internet users accessed the Internet from their smartphones (it should be noted that the region where the proportion of access to the Internet by mobile/smartphone is highest is Africa), statistics based on home or school access to the Internet have to be complemented by including the use of smartphones. Granted, however, that learning by smartphones offers, in theory, fewer possibilities than using a dedicated school computer. Yet, in the US in 2022, 55% of high school students used their smartphones for schoolwork. Since the technological affordances of smartphones grow exponentially, the real divide may be less in access to the Internet from home or school than access to broadband and quality (and price) of smartphone technology, therefore increasing the relevance of income, education, and infrastructure in the construction of digital inequality.

In sum, since inequality in the distribution of resources (be it economic or cultural) is prevalent around the world, and increasingly so (Piketty

& Rendall, 2022), it is not surprising that inequality is also a defining feature of the distribution of digital communication use and capacity at the global level. However, the question remains of the extent of digital inequality compared to other sources of inequality. There is scant quantitative research on the matter. Nonetheless, one indication in this regard comes from the study I directed on the diffusion, uses, and impact of mobile communication in Latin America (Fernández-Ardèvol et al., 2011). Comparing inequality between countries according to income and inequality according to the penetration of mobile communication, we observed that income inequality was more pronounced than mobile communication inequality. We ventured some hypothetical explanation for the reasons for the spread of mobile telephony throughout all of the social groups using ethnographic analysis. While higher-income groups obviously benefitted from better equipment and infrastructure, most low-income people (including the urban poor and farmers) felt the importance of using mobile communication, both for their working needs and their personal lives, including to keep emotional and financial ties with their family migrants.

Furthermore, a well-established literature has provided evidence of the correlation between the deployment of digital infrastructure and skills level, and the economic growth of countries (Vu et al., 2020). While it may appear that the most pressing needs are to remedy the lack of electricity, water, amenities, and public services, the capacity to finance those social investments depends on economic growth that, in the current global economy, is partly a function of the technological modernization of economies and societies. Ortiz Freuler has analyzed the relationship between inequality in countries and their level of Internet access, which shows a strong relationship between inequality levels as measured by the Gini index and Internet access between countries, where the lower the inequality, the higher the Internet access (see Figure 7.1). In-depth research analyzing 86 countries has found that results are varied across countries, but within low-income countries an increase in Internet access is associated with a reduction in poverty, and access to broadband with a reduction in income inequality (Afzal et al., 2022).

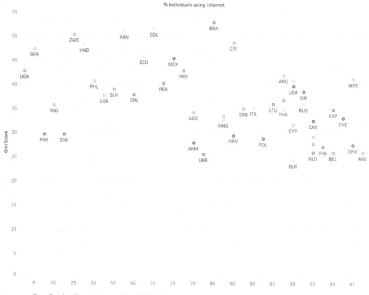

Source: By Ortiz Freuler using ITU data.

Figure 7.1   Scatterplot comparing Internet use and Gini score

Ethnicity divide

It is generally assumed that ethnic minorities are disadvantaged in terms of Internet access. We do not really know because many countries (France and Germany, for instance) do not include race or ethnicity in their Census categories, a dubious symbol of calling for citizens' equality beyond ethnicity. The US does have ethnically differentiated statistics, both in the Census and in a number of surveys (US Census Bureau, 2021). The story the data tell is nuanced. Counting the proportion of Internet users within each ethnic group, in 2019, the highest group, at 99%, were the Chinese-Americans. In 2021, a survey of adults of a representative sample of recent users of the Internet published by Pew showed some unexpected findings: while 93% of whites were users of the Internet, Hispanics had a higher proportion (95%) with African Americans not too far behind at 91% (Pew Research, 2021a). On the other hand, in 2021, there were significant differences in terms of home broadband: 80% of white homes have access, in contrast to 71% of African Americans, and

67% for Hispanics. One possible explanation for this discrepancy could be that the proportion of ownership of smartphones (with Internet access potential) was very similar: 85% for whites, 85% for Hispanics, and 83% for African Americans (Pew Research, 2021b). Much of the perceived gap between whites and other ethnic minorities results from differences in income, education, territory, and lack of infrastructure that disproportionately hit ethnic minorities. Rural African Americans, particularly in the Southern states, were nearly twice as often excluded than white Americans in the same areas (Asher-Schapiro & Sherfinski, 2021).

A new form of ethnic discrimination has appeared in the digital environment: some social media platforms allow their advertisers to discriminate against groups protected by civil rights laws. After the intervention of the Department of Justice, in 2019 Facebook settled a lawsuit by civil rights groups by creating a special portal for ads related to credit, housing, and employment, to ensure targeting techniques would not be used to discriminate against disadvantaged groups (Tobin & Kofman, 2022). This is probably only the tip of the iceberg of a new form of digital discrimination.

In sum, historically disadvantaged ethnic minorities suffer from a multiplicity of sources of inequality (income, assets, education, infrastructure) whose combined effect challenges the optimistic notion of equalization of opportunities in the digital age.

Cultural divides

We live in a multicultural world. And yet, while cultures are all born equal, history has not treated them equally, both within each country and in the global realm. Cultures are codified in language. Although language is not the only component of a given culture, it is the decisive one because it is what allows communication among those identifying with the culture. And yet, when we express ourselves and communicate in the global agora, which is what the Internet is supposed to be, most people have to use a language, thus a cultural construct, that is not theirs. Indeed, English is the Internet's universal language (Statista, 2022). According to UN statistics, about 1.5 billion people speak English (this of course includes part of the population of India), of which 1.2 billion are Internet users. This represents approximately 30% of the world's Internet users. A critical observation is that 70% of Internet users may speak English but are not native English speakers. This is largely due to Chinese, other Asians, Russians, Germans, Spanish speaking, Portuguese speaking, and Arabs, Iranians, and Africans that interact on the Internet within their language/cultural community. It is important to dispel the notion that communication on the Internet is in English. In its large majority it is not. But, global, intercultural communication is in English, as is, in fact, most international communication. The imbalance has consequences in the uneven access to resources on the World Wide Web. Figure 7.2 shows the relationship between the language of websites and the languages of Internet users, as of 2021/22.

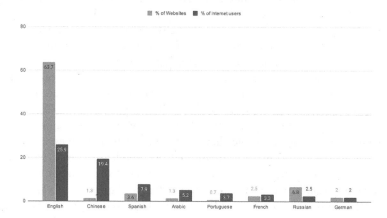

Note: Websites as of February 2022; Internet users as of 2021.
Source: Statista, W3CTechs, Internet World Stats.

Figure 7.2   Share of websites using selected languages compared to estimated share of Internet users speaking those languages

The gap becomes more significant when considering the relationship between the language distribution of the website and the share of Internet users speaking a given language. Overall, it means that 70% of users are unable to use their language on 60% of all websites. Since the knowledge of English is skewed toward the most privileged groups in every non-English-speaking country, a fundamental cultural divide is introduced in a digitized world.

Granted, substantial progress in automated translation operated by AI allows us to overcome some of the language barriers. But this creates an additional inequality in terms of the skills necessary to operate and interpret IA programs.

The latest developments on AI build upon LLMs trained on massive datasets containing the text collected from the open web. As a result, applications like ChatGPT perform better with languages for which more websites are available. ChatGPT operates on databases that are mainly in English. For example, although it is estimated that there are 7,000 spoken languages and approximately 300 writing systems (UNESCO, 2018), ChatGPT claims it can respond in English, Spanish, French, German, Chinese, Japanese, "and many more," but cannot establish a number to what others have estimated to be 95 languages (SEO AI, 2023). As a result, those who can speak popular languages will get access to state-of-the-art technologies, while others will be excluded from being able to take advantage of them. On the flipside, like the excluded proles of Orwell's *1984*, they might be the only ones left with the ability to engage online freely, while those whose languages have been studied by machines will find their expressions constantly monitored and moderated by automated systems. During the world wars, the US leveraged their unique indigenous languages as part of a codification system the Japanese and Germans would not be able to break (CIA, 2008). Perhaps in the cyberwars of the future, the languages unknown to the machines will play a similar role.

Furthermore, languages are produced by culture, and our brains use language categories for their codification of experience. Therefore, beyond language translation we have to engage in a cultural decoding of our global communication capabilities, emphasizing the fact that we live in a global world, but we do not have, and will not have, a homogeneous global identity. "Citizens of the world" is just the self-aggrandizing definition of the global gentry (Castells, 1997).

Meaningful access to the Internet

The United Nations 2030 Agenda proposes that Internet access should be universal and meaningful. The aim is 100% access for all adults, all schools, all businesses, with a mobile network of the latest technology, and universal ownership of a mobile phone by the adult population. Gender parity should be achieved, 70% of adults should have basic digital skills, 100% of fixed broadband subscriptions should be 10 Mb/s or faster, and 20 Mb/s should be the minimum download speed at every school. Entry-level broadband subscription should cost less than 2% of the average national income. The data presented in this chapter show how far most of the world still is from these targets. However, there is an even more fundamental question: what is the *meaning* of an Internet connection for people at large? A number of studies (Blank, 2013; Pew Research, 2013) have investigated the factors that lead people not to access the Internet. Old age is the most important factor. Indeed, in the US in 2021, 25% of people over 65 never go online. Low level in education deters potential users. Intensity of bonding and social capital also diminishes interest for Internet use, as does satisfaction with one's health condition. Gender, income, and occupation do not seem to predispose to this withdrawal from digital life. Thus, a potential portrait of the voluntary nonuser of the Internet emerges: individuals whose life cycle invites a retreat to their world of intimacy, and free from work-related pressures, most of them being retired; people in good health with strong social bonds who prefer selective person-to-person sociability, in other words, their meaningful communication is built outside the digital environment. On the opposite side of the spectrum, people with low education lack skills to find casual navigation of the Internet rewarding. Between the educationally excluded and those able to live life as it used to be, the overwhelming majority of the population needs digital networks to learn, work, and live, as citizens-netizens. It is out of the question not to use the Internet. So, they inhabit the Internet galaxy, where they meet the traditional sources of social inequality, and the uncertain quest for the meaning of their lives. The meaningfulness of the Internet does not result from any technological affordance. It depends on the meaning that everybody gives to their experience, thus adapting the new technologies to the fulfillment of their needs and desires.

References

Afzal, A., Firdousi, S.F., Waqar, A., & Awais, M. (2022) "The influence of Internet penetration on poverty and income inequality," *SAGE Open*, 12(3), 21582440221116104. https://doi.org/10.1177/21582440221116104.

Allcott, H., Braghieri, L., Eichmeyer, S., & Gentzkow, M. (2020) "The welfare effects of social media," *American Economic Review*, 110(3), pp. 629–676. https://doi.org/10.1257/aer.20190658.

Asher-Schapiro, A., & Sherfinski, D. (2021) "'Digital divide' hits rural Black Americans hardest," *Reuters*, 6 October. Available at: https://www.reuters.com/article/us-usa-internet-race-idUSKBN2GW1QQ (accessed on 2 September 2023).

Beneito-Montagut, R., Rosales, A., & Fernández-Ardèvol, M. (2022) "Emerging digital inequalities: A comparative study of older adults' smartphone use," *Social Media + Society*, 8(4), 20563051221138756. https://doi.org/10.1177/20563051221138756.

Blank, G. (2013) "Why have some people stopped using the Internet?," *OII*. Available at: https://www.oii.ox.ac.uk/news-events/news/why-have-some-people-stopped-using-the-internet (accessed on 16 September 2023).

Castano, C. (2005a) *Las mujeres y las tecnologías de información*. Madrid: Alianza.

Castano, C. (2005b) *La Segunda brecha digital*. Barcelona: Cátedra.

Castano, C. (2015) *Las mujeres en la Gran Recesión*. Barcelona: Cátedra.

Castano, C. (2021) *El papel de las mujeres en la revolución digital*. Madrid: Santillana.

Castano, C., & Webster, J. (2011) "Understanding women's presence in ICT: The life course approach," *International Journal of Gender, Science and Technology*, 3(2), pp. 364–386.

Castano, C., & Webster, J. (2014) *Género, Ciencia y Tecnologías de Información*. Barcelona: Aresta.

Castano, C. et al. (2010) *Las mujeres en las carreras de informática y telecomunicaciones*. Madrid: Ministerio de Industria, Research Monograph.

Castells, M. (1997) *The Power of Identity*. Malden, MA: John Wiley & Sons.

Chakravorti, B. (2017) "There's a gender gap in Internet usage. Closing it would open up opportunities for everyone," *Harvard Business Review*, 12 December. Available at: https://hbr.org/2017/12/theres-a-gender-gap-in-internet-usage-closing-it-would-open-up-opportunities-for-everyone (accessed on 16 September 2023).

CIA (2008) "Navajo code talkers and the unbreakable code." Available at: https://www.cia.gov/stories/story/navajo-code-talkers-and-the-unbreakable-code/ (accessed on 16 September 2023).

ESGN (2021) "China Mobile is bridging the gap between rural and urban digital divide in China," *ESGN Asia*, 5 July. Available at: https://esgn.asia/china-mobile-is-bridging-the-gap-between-rural-and-urban-digital-divide-in-china/ (accessed on 16 September 2023).

Fernández-Ardèvol, M. (2020) "Older people go mobile," in R. Ling, L. Fortunati, G. Goggin, S.S. Lim, & Y. Li (eds), *The Oxford Handbook of Mobile Communication and Society*. Oxford: Oxford University Press, pp. 187–199.

Fernández-Ardèvol, M., Castells, M., and Galperin, H. (eds) (2011) *Comunicacion movil y desarrollo economico y social en America Latina*. Barcelona: Planeta.

Fernández-Ardèvol, M., Sawchuk, K., & Grenier, L. (2017) "Maintaining connections: Octo- and nonagenarians on digital 'use and non-use'," *Nordicom Review*, 38(s1), pp. 39–51. https://doi.org/10.1515/nor-2017-0396.

Fernández-Ardèvol, M., Rosales, A., & Cortès, F.M. (2023) "Set in stone? Mobile practices evolution in later life," *Media and Communication*, 11(3), pp. 40–52. https://doi.org/10.17645/mac.v11i3.6701.

Fink, E., Patalay, P., Sharpe, H., Holley, S., Deighton, J., & Wolpert, M. (2015) "Mental health difficulties in early adolescence: A comparison of two cross-sectional studies in England from 2009 to 2014," *Journal of Adolescent Health*, 56(5), pp. 502–507. https://doi.org/10.1016/j.jadohealth.2015.01.023.

Galperin, H., Katz, R., & Valencia, R. (2022) "The impact of broadband on poverty reduction in rural Ecuador," *Telematics and Informatics*, 75, 101905. https://doi.org/10.1016/j.tele.2022.101905.

Garín-Muñoz, T., Pérez-Amaral, T., & Valarezo, Á. (2022) "Evolution of the Internet gender gaps in Spain and effects of the Covid-19 pandemic," *Telecommunications Policy*, 46(8), 102371. https://doi.org/10.1016/j.telpol.2022.102371.

ILO (2019) "Tech's persistent gender gap," *ILOSTAT*. Available at: https://ilostat.ilo.org/techs-persistent-gender-gap/ (accessed on 16 September 2023).

Mubarak, F., Suomi, R., & Kantola, S.-P. (2020) "Confirming the links between socio-economic variables and digitalization worldwide: The unsettled debate on digital divide," *Journal of Information, Communication & Ethics in Society*, 18(3), pp. 415–430. https://doi.org/10.1108/JICES-02-2019-0021.

NCES (2021) "Students' Internet access before and during the coronavirus pandemic by household socioeconomic status," *NCES Blog*. Available at: https://nces.ed.gov/blogs/nces/post/students-internet-access-before-and-during-the-coronavirus-pandemic-by-household-socioeconomic-status (accessed on 24 September 2023).

OECD (2018) "Empowering women in the digital age." Paris: OECD. Available at: https://www.oecd.org/social/empowering-women-in-the-digital-age-brochure.pdf (accessed on 16 September 2023).

ONS (2021) "Internet users, UK." *Office for National Statistics*. Available at: https://www.ons.gov.uk/businessindustryandtrade/itandinternetindustry/bulletins/internetusers/2020 (accessed on 26 September 2023).

Ortiz Freuler, J.O. (2020) "The shape of the Internet: A tale of power & money," *Medium*. Available at: https://juanof.medium.com/the-shape-of-the-internet-a-tale-of-power-money-a08d01065bc0 (accessed on 16 September 2023).

Parzyan, A. (2023) "China's Digital Silk Road," in M. Sahakyan (ed.), *China and Eurasian Powers in a Multipolar World Order 2.0: Security, Diplomacy, Economy and Cyberspace*. London: Routledge.

Pew Research (2013) "Who's not online and why," *Pew Research Center: Internet, Science & Tech*, 25 September. Available at: https://www.pewresearch.org/internet/2013/09/25/whos-not-online-and-why/ (accessed on 16 September 2023).

Pew Research (2021a) "Internet use by race/ethnicity," *Pew Research Center: Internet, Science & Tech*. Available at: https://www.pewresearch.org/internet/chart/internet-use-by-race/ (accessed on 16 September 2023).

Pew Research (2021b) "Mobile fact sheet," *Pew Research Center: Internet, Science & Tech*. Available at: https://www.pewresearch.org/internet/fact-sheet/mobile/ (accessed on 16 September 2023).

Pew Research (2022) "Teens and cyberbullying 2022," *Pew Research Center: Internet, Science & Tech*, 15 December. Available at: https://www.pewresearch.org/internet/2022/12/15/teens-and-cyberbullying-2022/ (accessed on 26 September 2023).

Piketty, T., & Rendall, S. (2022) *A Brief History of Equality*. Cambridge, MA: Harvard University Press.

Rajam, V., Reddy, A.B., & Banerjee, S. (2021) "Explaining caste-based digital divide in India," *Telematics and Informatics*, 65(C), 101719. https://doi.org/10.1016/j.tele.2021.101719.

Satariano, A. (2019) "How the Internet travels across oceans," *The New York Times*, 10 March. Available at: https://www.nytimes.com/interactive/2019/03/10/technology/internet-cables-oceans.html (accessed on 11 September 2023).

SEO AI (2023) "How many languages does ChatGPT support? The complete ChatGPT language list." Available at: https://seo.ai/blog/how-many-languages-does-chatgpt-support (accessed on 16 September 2023).

Shensa, A. Sidani, J.E., Dew, M.A., Escobar-Viera, C.G., & Primack, B.A. (2018) "Social media use and depression and anxiety symptoms: A cluster analysis," *American Journal of Health Behavior*, 42(2), pp. 116–128. https://doi.org/10.5993/AJHB.42.2.11.

Statista (2020a) "United Kingdom (UK): rural household Internet access 2011–2020," *Statista*. Available at: https://www.statista.com/statistics/1236330/share-rural-households-internet-access-united-kingdom/ (accessed on 26 September 2023).

Statista (2020b) "United Kingdom (UK): urban household Internet access 2020," *Statista*. Available at: https://www.statista.com/statistics/1235783/internet-access-urban-households-united-kingdom-uk/ (accessed on 26 September 2023).

Statista (2021) "Global Internet users age distribution 2021," *Statista*. Available at: https://www.statista.com/statistics/272365/age-distribution-of-internet-users-worldwide/ (accessed on 16 September 2023).

Statista (2022) "Infographic: English is the Internet's universal language," *Statista Infographics*. Available at: https://www.statista.com/chart/26884/languages-on-the-internet (accessed on 9 April 2023).

Statista (2023) "5G network availability by country 2023," *Statista*. Available at: https://www.statista.com/statistics/1215456/5g-cities-by-country/ (accessed on 16 September 2023).

Telegeography (2018) *Global Internet Map 2018*. Available at: https://global-internet-map-2018.telegeography.com/ (accessed on 16 September 2023).

Tobin, A., & Kofman, A. (2022) "Facebook finally agrees to eliminate tool that enabled discriminatory advertising," *ProPublica*. Available at: https://www.propublica.org/article/facebook-doj-advertising-discrimination-settlement (accessed on 2 September 2023).

Twenge, J.M. (2023) *Generations*. New York: Atria Books.

UNESCO (2018) *The World Atlas of Languages*. Available at: https://en.wal
.unesco.org/world-atlas-languages (accessed on 16 September 2023).

US Census Bureau (2021) "Computer and Internet use data," *Census.gov*. Available
at: https://www.census.gov/topics/population/computer-internet/data.html
(accessed on 16 September 2023).

Vu, K., Hanafizadeh, P., & Bohlin, E. (2020) "ICT as a driver of economic
growth: A survey of the literature and directions for future research,"
Telecommunications Policy, 44(2), 101922. https://doi.org/10.1016/j.telpol
.2020.101922.

Wetsman, N. (2021) "Facebook's whistleblower report confirms what researchers have known for years," *The Verge*, 6 October. Available at: https://www
.theverge.com/2021/10/6/22712927/facebook-instagram-teen-mental-health
-research (accessed on 26 September 2023).

Ziegler, S., Arias Segura, J., Bosio, M., & Camacho, K. (2020) "Conectividad rural
en América Latina y el Caribe. Un puente al desarrollo sostenible en tiempos
de pandemia." *Instituto Interamericano de Cooperación para la Agricultura
(IICA)*. Available at: https://repositorio.iica.int/handle/11324/12896 (accessed
on 11 September 2023).

8 Networked social movements

Social movements have always been the primary agents of social change. They aim at transforming what we think and what we value. Thus, they are cultural movements rather than political movements, those seeking to alter power relations in the State, although they often induce political effects as a result of the new values they put forward. They are formed by communication between people challenging the dominant norms enforced by institutions of society. Frequently they result from outrage provoked by acts of powerful actors that are perceived as unjust. Social mobilization follows as a protest against those acts. From these collective actions emerge ideas that are elaborated and debated in the movement, and ultimately may induce projects of social transformation, whose fate depends on the conflictive interaction between social movements and the institutions of society. Some movements become formal organizations. Most of them exist as loose forms of cooperation structured and focused in a communication process. How communication is established and which are the technologies of communication evolve historically and ultimately frame the social movements. In the twenty-first century, the advent of digital networked communication, as analyzed in Chapter 2, has induced a form of social movements characteristic of our time: networked social movements, whose features and dynamics I will present in this chapter. Networked social movements are not the result of digital communication. They respond to perceived oppression and injustice that are rooted in the dominant values and interests in a given society. Yet, the form they take, and their practices, are enabled by digital communication technologies. They evolve with technological change in interaction with cultural and political change.

The early twenty-first century: a time of global change[1]

The sudden rise of networked social movements took place in the second decade of the twenty-first century, particularly after 2010, although before this date there were a number of powerful mobilizations facilitated by digital mobile phones in countries as different as South Korea (2002), Spain (2004), and Iran (2009).

However, the spread of these movements picked up in 2010 and subsequent years. In December 2010 similar movements first started on the Internet and then occupying urban space took place in countries as different as Iceland and Tunisia. The Tunisian revolution sparked what was labeled the Arab Spring in January 2011, with hundreds of thousands of people following calls on the Internet, motivated by outrageous behavior of the ruling elites. They occupied squares in the main cities in Egypt (Cairo's Tahrir Square became a symbol all over the world), Syria, Bahrain, Morocco, and smaller movements in other Arab countries. In Spain, the May-15th 2011 movement, protesting the management of the financial crisis and calling for "real democracy," started on the Internet and went on to occupy squares for over a month in most Spanish cities. Similar movements, with diverse intensity, took place in other European cities: Paris Debout, Lisbon, Athens, London, Amsterdam, Berlin. In Istanbul the Gezi Park movement challenged the municipal authorities. In Kiev, thousands of protesters opposed to the pro-Russian Government occupied Maidan square and prompted the Orange revolution in Ukraine, leading, years later, to a major impact on global geopolitics.

In the US, the Occupy Wall Street movement, started in September 2011, denounced the practices of financial speculation, and the protection of corporations by the government using public funds at a time when thousands of families were losing their homes because of mortgage defaults. The movement made the country aware of extreme inequality, with

---

[1] Detailed empirical studies and sources on networked social movements 2010–15 are presented in Castells (2015). Subsequent Internet-based social movements in 2016–23 as recorded by academic research and limited ethnographic observation by this author have reproduced a similar pattern in their social practice, confirming the main lines of the analysis presented in this chapter.

1% of people controlling most of the wealth. "We are the 99%" was the slogan of the movement. From New York, occupations extended to Los Angeles, Oakland, Chicago, Atlanta, and to over 1,000 American cities with variable intensity. In Brazil, in June 2013 the youth movement for free transportation triggered a series of demonstrations against political corruption that shook up the Brazilian institutions. New movements in 2015 and 2016 showed that the elites had learned the new form of politics and mobilized the middle class to elect as president the ultrarightist Bolsonaro. However, in 2022, left-wing activists regained influence in the Brazilian digital networks and succeeded in just electing Lula, ousting Bolsonaro. In Chile, the student movement that started in 2012 was able to mobilize other segments of the population, leading in October 2019 to a series of occupations and demonstrations in defense of Dignity. As a direct outcome of this movement, in November 2021, Gabriel Boric, a young leader of the movement, was elected president. In Colombia, in April 2021, the "estallido social" (social explosion) took place against the mishandling of the health measures against the Covid-19 pandemic, and against murderous police violence on the demonstrators. In 2022, largely as a result of the "estallido social," left-wing leader Gustavo Petro was elected president of Colombia and engaged in a determined peace process by negotiating with the guerrillas. In Mexico, social movements in the Internet, such as "#Yo soy 132" in 2012, were paramount in the defense of human rights and the exposure of corruption in the State. It paved the way for the election of populist president López Obrador. Less followed by the Western media, there were significant movements in other countries, in fact in over 100 countries. In Nigeria in 2020, the "End SARS" movement mobilized the population against a brutal special police unit. The government repressed the demonstrators and closed Twitter for 7 months. as social networks had played a key role in mobilization. The youth activists of this movement came very close to electing president Peter Obi (Labour Party) in the 2023 election, although the usual political tricks derailed his probable victory.

In Hong Kong in 2014, the Umbrella Movement fought for democracy and was able to call massive demonstrations in spite of heavy-handed police repression. Even in China, a country where these movements had not appeared until then, in 2022 there were significant protests organized on the Internet in Shanghai and other cities against the confinement policies of the government during the resurgence of the Covid-19 pandemic.

In fact, these movements were able to reverse government policies, a rare case in China

In Iran in 2022, thousands of Iranian women, connected around mobile social networks, and supported by men, protested against the assassination of Mahsa Amini at the hands of the religious police. The movement was savagely repressed and some activists were hanged. Yet, the seeds of protest and the affirmation of new values remain in the minds of Iranian women. In Israel in 2023, hundreds of thousands marched against the violation of the constitution by the Netanyahu government. In France in June 2023, the French "banlieues," where ethnic minorities are segregated on the periphery of major cities, revolted violently for over one week, protesting the assassination by the police of 17-year-old Nahel, and calling for an end to institutional racism, prevalent in the police.

Thus, networked social movements cut across ideologies, issues, or institutions, as they become the global form of social change in our time.

Anatomy of networked social movements

Networked social movements are highly diverse around the world. Yet, they share some fundamental characteristics in spite of substantial differences between the cultural and institutional contexts in which they happen. Indeed, it is because of this similarity in spite of contextual differences that I suggest that we are in the presence of a distinctive form of social movement that emerges from the organizational features and communication technologies underlying our social structure on a global scale: the network society (Castells, 2000). Observation of these social movements by a plurality of researchers, including myself, shows a recurrent pattern, with slight variations depending on specific settings (van de Donk et al., 2004; Earl & Kimpost, 2011; Castells, 2015; Carty, 2018; Jackson, 2018; Brown et al., 2022). I will summarize below the main components of this common pattern and the conditions of its interaction with the social structure and institutions where collective action aimed at social change takes place in contemporary societies.

First of all, their organizational form and their practices are based on networks: social networks and communication networks usually enabled by digital communication technologies, particularly mobile communication devices connected by the Internet. Since the smartphone was first introduced in 2007, it appears probable that this technology made possible an exponential increase of social networks in the practice of social movements at the same time that the networking logic diffused in all domains of human communication and organization. Indeed, while mobile phones played a significant role prior to 2007 in some social protests, the growth of Internet-based communication networks allowed the explosion of these movements after 2010.

Second, although the large majority of these movements formed through interactions on the Internet, they became socially meaningful by networking people in the urban space, be it in street demonstrations, as in Brazil, Syria, and Iran, or by occupying squares and buildings for a period of time, as in Egypt, Spain, the US, and Ukraine. The spatial expression of the movement, regardless of its duration in time, is essential to make it visible to the entire society, beyond those connected on the Internet. Furthermore, the physical connection enhances the feeling of togetherness that is a source of strength necessary to face the fierce police repression that will inevitably hit the movement. In fact, the simple fact of illegally occupying space challenges the powers that be, and materializes the protest in a form that makes it easy for people to join the movement, just by going to the occupied space without requiring a formal membership.

Thirdly, the combination of networked interaction both in cyberspace and in urban space constructs materially and culturally a space of autonomy, an essential feature of social movements in all historical instances, as this autonomy is required to freely socialize, deliberate, and decide outside the reach of the established procedures of the society that are being challenged. Looking back in history, the industrial working-class movement in Britain emerged from the togetherness felt in the large factories, in combination with spaces of autonomy such as the British pubs (Thompson, 1963). The Commune of Paris in 1871 and multiple movements of alternative projects of common life, in the utopian or anarchist traditions, played a similar role (Castells, 1983). The space of autonomy in the network society is built around the Internet networks and in the occupation of urban space, however brief this may be.

This form of autonomy explains the fourth and most essential feature of these movements: they are rhizomatic. When the institutions of society (judicial authorities, police, and the like) force the movement to retreat, overwhelmed by coercion, more often than not they do not disappear: their values, their projects, as well as many of their actors take refuge in the Internet, and they surge again when the reproduction of social injustice at the roots of the outrage that gave birth to the movement recur. The . rhizomatic feature of these movements helps their survivability. While Internet networks can be surveilled, and some rare times even shut down, our societies cannot operate without the Internet and digital communication networks, and so any dormant seed of resistance can be activated at any time, although with different themes and projects.

Fifth, networked movements do not need formal organization or established leadership. The network is the message. Its nodes and their relative hierarchy evolve with the movement and its interaction with the institutions. Membership is loose, discipline absent, replaced by influence and self-organization. When the movement evolves to involve political actors with specific programs, then formal organization and leadership become essential. But this is a different kind of social actor whose origins may be in the social movements but whose logic and practice are substantially different. They are political actors, seeking political power by acting on the State. Social movements, in contrast, explore and propose values, blueprints of life, acting on the minds of the people.

Lastly, social movements are rarely programmatic. They put forward demands and refer to values and principles, but they do not propose a realistic set of institutional measures. In fact, in many cases the movement anticipates in its practice what would constitute its desirable forms of institutional organization. For instance, the frequent goal of achieving "real democracy" is exemplified in the practice of deliberative democracy, debating and deciding in open assemblies in the squares and in consultations on the Internet. In other instances, social revolts propose values in their practice without making them explicit. For instance, the Black Lives Matter movement in the US in 2021, or the violent protests in the French "banlieues" in 2023, shifted from denouncing police profiling and unchecked brutality to the ideal of eradicating racism from the institutions. This is a fundamental value to ensure peaceful coexistence in our multiethnic societies that is implicit in the insurgency against the systemic police repression that targets ethnic minorities. To end racism would be a major cultural and political revolution. However, while movements are not programmatic *stricto sensu*, there are certain values that appear consistently in the discourse of many movements. The most important of these common values, across cultures and countries, is the defense of Dignity, a term repeated in different languages, from Madrid to New York, and from Kiev to Santiago de Chile. People mobilized to be recognized and respected as human beings, regardless of their gender, age, ethnicity, education, income, or legal status. It would appear that beyond the specific sources of the protest and the goals of the mobilization, most of these actions are movements for human rights, encompassing all the claims for equal treatment to people by the institutions beyond the narrow definition of citizen rights. Dignity is the value claimed over the entire world by movements that aspire to expand equality and liberty to all domains of human experience.

Social movements always fade away, either obliterated by repression (including cultural repression in and by the mass media), or co-opted and institutionalized in their values and in their actors. Yet, the forms of their death are not irrelevant. They may die in an apparently useless effort to erode the fortresses of society. But also, in many cases, they survive and blossom in terms of the ideas and ideals that they project, out of the outrage that prompted their existence. They can induce hope for a better life and a better world in the minds of people, ultimately enshrined in the institutions. This is why they are so deeply misunderstood by media and politicians, always obsessed with tangible results. In fact, their potential outcome is the transformation of values and institutions of society. They are the salt of the Earth.

Social movements and communication technologies

Social movements are a permanent feature of human history. They display multiple forms, but are always based on a process of communication by which outrage is shared, demands and values bring people together, and collective action is organized. This is why the control of information and communication was always paramount to preserve power. Consequently, only by breaking the hold of the powerful elites over communication could social change occur. This could be the work of prophets, of oral diffusion of criticism and new ideas, of communicable texts, and ultimately of pamphlets, books, and the print press. This is why books were banned or burned, and reading and writing was confined to the elites for centuries, and censorship, by religious or political authorities, was and is a decisive lever of power making (Castells, 2009). With the advent of digital social networks communicated over the Internet, the monopoly of mass communication was challenged. Not that power dissolves, because it is present in the interaction in the social networks, and because corporations own the networks and governments retain regulatory powers, as I have analyzed in previous chapters of this volume. Yet, full control of information becomes more difficult, and the communicating actors have a fighting chance to connect, share outrage, counteract manipulative messages, and elaborate alternative projects.

Social protests, eventually developing into social movements, are not the result of digital networks of communication. They surge from the pits of suffering of everyday life. But because our society is largely based on these networks, in every domain, their logic and their affordances extend to all human practices, including social movements. This is why in the network society social movements take the form of networked social movements. They could not exist in their current form without the digital communication paradigm. Technological foundations of collective action do not prejudge the values or projects of the networked actors. Social mobilizations aiming at the restoration of the traditional social order coexist with movements seeking social change. In both cases their logic is shaped by the specific characteristics of digital technologies.

Digital communication is based on networks, and this is the form that practically all social protests and social movements take in our time. What follows is the increasing survivability of networked social movements, in terms of their rhizomatic dynamics and their resilience to targeted repression. They often become instant "communities of practice," without the need for formal organization and leadership, because their coordination and deliberation can always happen over the Internet. Communities of practice are increasingly important forms of social organization in all domains of society, from business to science and education (Hughes et al., 2013). They also characterize social movements, transforming their dynamics and making possible horizontal patterns of coordination and guidance.

Outrage at an action by a powerful social actor has always been the most frequent stimulus to spontaneous protests that often overwhelm the repressive capacity of the established institutions, forcing at least a negotiation to preserve social order under new regulations. Shared outrage continues to be the spark that lights the fire of social revolts in our society. The specificity of our contexts is the capacity to visualize and distribute images of the outrageous behavior. The influence of images in our minds is the most important trigger of emotions. And emotions are the main determinant of collective action. YouTube, Instagram, Twitter, and TikTok, among other social networks, have been the most decisive tools to ignite social protests. Furthermore, the pervasiveness of smartphones makes it possible for anyone to record an outrageous act (such as an arbitrary killing by a police officer or the torture of animals) and immediately upload it to the Internet.

Virality is a fundamental feature of Internet communication (Nahon & Hemsley, 2013). Virality makes possible the almost immediate distribution of a message to a very large audience. The quasi-simultaneity of the reception of the message helps the spontaneity of the protest, by easy, fast feedback between communicating subjects who, in their interaction, scale up from individual emotion to collective mobilization. The new technologies of connectivity (5G, 6G), besides increasing the speed of communication, reduce the latency of the response, facilitating the interaction of outraged individuals in real time.

In sum, digital communication technologies, combined with imaging techniques, and the uses of artificial intelligence in constructing images, shape social movements caused by reactions against perceived social injustice in ways that empower considerably the new actors of the process of social change.

Networked social movements and political change

While the fundamental outcome of social movements refers to changes in our way of thinking and being, with different timing depending on contexts, the most immediate traces of their potential as agents of change appear to be in their impact on the State, as the direct expression of power relations. The empirical record on this issue shows a wide range of outcomes. In most cases of the observed practices in the second decade of the twenty-first century, the political outcomes have been highly significant. However, not necessarily by fulfilling the goals of the movement. In fact, in many instances the powers that be reacted with extreme brutality and crushed the movements, sending dark waves of despair across the world. The Arab Spring did topple decades-old dictatorships in Tunisia and Egypt, and challenged Assad in Syria. Yet, Tunisia ultimately became a corrupt pseudo democracy, the Army took control over Egypt, supported by the US, and Russia came to the rescue of Assad. Then, Al Qaeda, Saudi Arabia, Iran, Israel, and the Western powers joined the fray, triggering an atrocious war in Syria with hundreds of thousands of deaths and millions of refugees.

Social movements disappeared because they are always the first victims of wars. Ideas may prevail in the long term. But missiles, tanks, and terror overwhelm ideals in the short term. Nonetheless, I dare to consider these catastrophic political effects as being induced by the social movements of the Arab Spring. Negative effects are, analytically speaking, as important as positive ones. In the US, the Occupy Wall Street movements planted seeds for the crisis of elite legitimacy in the minds of millions of citizens. For instance, they became fully aware of the extent of social inequality contradicting the American dream. Institutions were delegitimized, including political parties, financial elites, and the media. Paradoxically, the main political outcome was the rise of right-wing populism, exemplified by Trump and his Make America Great Again (MAGA) ideology reaching the presidency against all odds, thus changing American politics and ultimately endangering democracy. This is because the challenges that emerge from social movements have to be processed in the political system, and when there is no relay between alternative projects for society and the political actors, opportunistic demagogues tap on the nostalgia of better times that never were, instead of exploring new forms of living together.

However, in several countries some social reforms and political democratization were implemented as a partial response to social movements. In Latin America, social movements resulted in major progressive political changes in Chile, Colombia, and Mexico, in spite of the limitations of left-wing populism. Spain offered perhaps the most direct example of how powerful social movements led to the emergence of a new political actor (Podemos) and to a new left-wing orientation of the powerful Social

Democratic Party. Together they formed the only center-left coalition government in Europe, coming to power in 2020, at the very time that the Covid-19 pandemic hit the world and the country. They successfully managed the health situation and the economy, and they enacted a number of progressive environmental, feminist, and social policies. Only to be hit again, in 2022, by the Ukraine war and the war-induced inflation. Yet, they were reelected in 2023 in alliance with Catalan, Basque, and Galician nationalists, making possible a multinational coalition in the Spanish State for the first time in history.

This is to say that social movements, by bringing new values to the public agenda, may change the political scene and government policies, but political outcomes will depend on the interplay between political actors and political dynamics in society at large. And societies are usually prisoners of fear and averse to profound changes. So, from outrage to hope to political change, there are a number of intermediate institutional layers, specific to each society, that refract and modify any new project of life created by the social movements. In this interplay between social change and social resistance to change, communication networks are essential as they process the ideas and will that configure the minds of citizens. Networked social movements interact with networked informational politics (Castells, 2017).

References

Brown, N., Block, R., & Stout, C. (2022) *The Politics of Protest: Readings on the Black Lives Matter Movement*. London: Routledge.

Carty, V. (2018) *Social Movements and New Technology*. London: Taylor & Francis

Castells, M. (1983) *The City and the Grassroots*. Berkeley, CA: University of California Press.

Castells, M. (2000) *The Rise of the Network Society*. Oxford: Wiley-Blackwell.

Castells, M. (2009) *Communication Power*. Oxford: Oxford University Press.

Castells, M. (2015) *Networks of Outrage and Hope: Social Movements in the Internet Age*. Cambridge: Polity.

Castells, M. (2017) *Rupture: The Crisis of Liberal Democracy*. Cambridge: Polity.

Earl, J., & Kimpost, K. (2011) *Digitally Enabled Social Activism*. Cambridge, MA: MIT Press.

Hughes, J., Jewson, N., & Unwin, L. (2013) *Communities of Practice: Critical Perspectives*. London: Routledge.

Jackson, S. (2018) "Progressive social movements and the Internet," Research paper, Philadelphia, Annenberg School of Communication, University of Pennsylvania.

Nahon, K., & Hemsley, J. (2013) *Going Viral*. Cambridge: Polity.

Thompson, E.P. (1963) *The Making of the English Working Class*. London: Penguin Books.

van de Donk, W., Loader, B.D., Nixon, P.G., & Rucht, D. (eds) (2004) *Cyberprotest: New Media, Citizens and Social Movements*. London: Routledge.

9 Social media and political polarization[1]

In all societies and throughout history, power relationships are the producers of norms and institutions of governance. Power relationships are by and large played out in the realm of communication (Castells, 2009). This is because the neural networks of human brains communicate with the brains of other humans and with the networks of their natural and social environment. Humans are social animals; thus, they live and act by their communication networks (Lakoff, 2008). From this interaction results the forms of social organization and institutions in which power relationships are embedded. The stable reproduction of these relationships takes place in the political process, which frames the competition for power positions in society. Institutional stability depends on the acquiescence of subjects of a given social system to respect the interests and values of actors who occupy power positions in the institutions. In this case, the institutional system is perceived as legitimate, or inevitable, by a sufficient majority of citizens. Yet, all institutions are in flux as actors, values, and interests that are in a subordinate position try to improve their standing by negotiating or challenging the norms enforced by the State. When the challenge from social actors reaches a certain level of intensity, the political institutions and their smooth operation suffer a crisis of legitimacy. Namely, large segments of society reject their authority.

A key factor in this potential crisis of legitimacy is the social dynamics taking place in the space of mass communication—a space that, as I explained in Chapter 2, depends on the characteristics of the senders of messages that could reach a mass audience, and on the technological medium they use. This space is organized and regulated, and becomes a critical component for the assertion of political legitimacy as well as for its potential crisis.

---

[1] The analysis presented in this chapter does not apply to the use of social media in authoritarian regimes, as polarization is suppressed by political censorship.

For most of human history, the technology of mass communication made possible the formation of organizations and institutions that were in control of the communication process, be it by the State, by economic organizations (e.g., corporate business), or by ideological apparatuses (e.g., churches or religious institutions at large). With the rise of networked digital communication technologies, this control is weakened. People-to-people networks can now reach an interactive, horizontal form of mass communication, disintermediating the powers that be. However, this techno-utopia is somewhat belied by the appropriation of these new platforms of digital communication by the new forms of capital and the State, the old nemeses of unfettered communication.

Yet, since communication controls are more difficult to enact in the digital paradigm, and since the new masters of communication have vested interests in stimulating communication within their proprietary networks, the scale and diversity of horizontal communications is such that relative communicative autonomy may challenge the systemic domination of established values and interests. This is why the traditional political institutions usually blame the vitality and diversity of social media communication, in spite of their oligopolistic appropriation, as the source of the crisis of legitimacy they suffer all around the world. In their view, this new trend results in an increasing ideological and political polarization that shakes institutional stability, ultimately threatening the very survival of liberal democracy.

Society *is* increasingly polarized. This polarization occurs in multiple domains of social life as major conflicts shake out the common ground of beliefs. This can be seen in the prevalence of sexism, homophobia, racism, xenophobia, religious fanaticism, radical nationalism, ethnic hatred, according to data from reliable sources (Achiume, 2018; OHCHR, 2019; Khan, 2021; World Health Organization, 2021; Sardinha et al., 2022).

Each one of the expressions of nonnegotiable antagonism has specific historical, cultural, and social roots. But what characterizes the current polarization is the fading away of trust in the institutions and leaders of all countries. Without trusted mediating institutions able to manage conflicts, the clash in values and interests appears to be irreconcilable. The end of trust ushers in an era of uncertainty. But what is the specific role of social media in the process of polarization?

A stream of scholarly research in recent years shows a complex picture of the ideological transformation underway. There is no doubt that the legitimacy of established institutions has been substantially eroded in the twenty-first century (Castells, 2009, 2015, 2017), not just concerning the State, the political system, and the political actors. It also refers to the mainstream media, the main economic actors (particularly the financial system) as well as traditional religions, official culture, and even science— as apparent by significant resistance to the vaccines that saved our lives during the Covid-19 pandemic.

In my book *Rupture: The Crisis of Liberal Democracy* (2017), I documented the extent of the legitimacy crisis worldwide, starting with the lack of trust of two-thirds or more of citizens in political parties and political leaders that govern their societies. Distrust of the elites and of their messages has dramatically increased since then in democratic countries (Edelman, 2022), partly as a consequence of the anger provoked by the confinement measures imposed to control the pandemic. Key factors in inducing the crisis of legitimacy appear to be the feeling of unfairness derived from rampant social and economic inequality, the resentment of government's systemic support of oligopolistic corporations, the widespread bureaucratization and perceived corruption of political parties and their leaders, and the shrinkage of the welfare state after the policies of austerity that followed the 2008–12 financial crisis.

Other factors contribute to the downfall of public trust: fear of fast, uncontrolled technological change, the backlash against the challenge to patriarchal domination, and their corollary: pervasive violence in women's everyday lives. Moreover, the growing awareness that climate change threatens the survivability of humans on the blue planet deepens uncertainty about our future.

Altogether, a perfect storm is forming with nobody at the helm of our fragile vessel of survival because of the distrust of institutions and elites. Collective uncertainty and individual anxiety populate the Internet, spreading rumors, disinformation, and misinformation under multiple disguises. We live in a post-truth age, and a key component of this misinformed bewilderment is the distortion of reality in a mass self-communication system with little regulation. Because the mass self-communication sphere that comprises billions of daily users is largely in the hands of private oligopolies whose business model is based on increasing traffic and generating data that they can capture with little government or judicial control (Taplin, 2023). Data capitalism is one of the most dynamic sectors of informational capitalism, as I analyzed in Chapter 3. The outcome of such a multidimensional transformation of communication is an immense cacophony of segmented, mass communication that mixes truth, lies, and partial lies, in a cloud of special effects and symbolic manipulation operating at light speed, saturating our lifetime, colonizing our personal space, thus blurring time, space, and experience.

Yet there is something else: is our daily practice of social media and their pervasiveness the primary source of the ideological and political polarization that is making it increasingly difficult to live together (Touraine, 2012)? These are the kinds of questions for which social science could provide some answers without yielding to the temptation of just asserting a subjective preference. The answer, that I will try to support empirically, is *yes* and *no* at the same time.

The bulk of political and ideological expressions in social media chat and publishing outlets are definitely polarized. However, what we would normally classify as political and ideological exchanges are relatively few in number when considering the scale of people participating within these sites. True, anyone with an opinion can post practically anything without yielding to any social norms of respect vis-à-vis dissenting opinions (Persily & Tucker, 2020).

Furthermore, most people gravitate to social networking sites, which includes specific groups within a platform, that reflect their own views, and toward mass media broadcasters that will confirm their views. This is because our brains tend to discard information that challenges any deep-seated perceptions and opinions (Castells, 2009: 167–169). A number of studies have shown, for instance, that, in the US, the audiences of Fox News and of National Public Radio do not overlap at all (Arsenault & Castells, 2008). In fact, the business model of Fox News aims at controlling the audience of ultraconservative viewers. The same logic applies within social media.[2] Groups within social media platforms are formed by affective affinity (Bakshy et al., 2015). The formation of

---

[2] It is interesting to look at the ideology of people who financially supported Elon Musk in his bid to take over Twitter, a company playing a role in

attitudes and the projection of behavior take place in what is known as "echo chambers," where the interacting subjects are rewarded by confirming their opinions and elaborating their views on a shared value system (Barbera, 2020). Thus, the polarization that exists in society determines the polarization in the mass media and in the media.

What is the role of media interaction with these pre-established opinions? It depends on how social media sites construct the interaction. The seminal work by Tucker et al., analyzing the empirical evidence on the matter, highlights three fundamental findings (2018):

1. Increased media fragmentation that mirrors the cultural and political fragmentation of society, replacing political news with entertainment, which ultimately lowers the quality of the political information received in the media space.
2. On the other hand, those people who do consume political information in the social media at large are exposed to a greater diversity of opinions, which may increase their civic engagement.
3. The exchange of explicitly political views in social media sites is frequently negative and uncivil, contributing to affective political polarization.

Therefore, the use of social media for all kinds of purposes may in fact increase civic consciousness and tolerance, while participation in partisan social media sites and conversations exacerbates polarization. Barbera, discussing a substantial body of research on the matter, finds that: "It is criticism of partisan identities, and not necessarily online opinions on specific issues that drives polarization" (2020: 45). In this regard, I would like to point out a crucial observation that has influenced the conclusions of many researchers: according to the study by Boxell et al. (2017) "greater Internet use is not associated with faster growth of political polarization among US demographic groups." Indeed, while between 1996 and 2016 political polarization increased substantially in all age groups, it increased twice as much among people older than 65, than among those aged 18–39, while the younger population is the most active group in using the Internet. As they write: "We find that polarization has increased the

---

a sector where most CEOs and employees are more aligned with the values professed by the Democratic Party.

most among the groups least likely to use the Internet and social media" (p. 10612).

However, for people who are already politically polarized, the use of partisan platforms and social media groups exacerbates their radicalization, particularly among those on the extreme right. It deepens polarization and intolerance in society at large because they are the most visible in the communication space (Nielsen & Fletcher, 2020). Four mechanisms seem to be operating to induce the polarization effect. The first is the addictive quality of intense engagement in social media controversies. As mentioned in Chapter 2, social media platforms are designed to keep users on their platforms for as long as possible. The timing of the delivery of "likes" and comments from followers and users sharing the same views, as well as the negative reaction to comments from opposite views, is leveraged and manipulated to produce a rush of dopamine, rewarding the brain with a feeling of excitement and satisfaction. Thus, the more people are engaged in social media activity, the more the dopamine neurons are active. And the anticipation of pleasurable effects increases participation in the social media that support this activity in a recurrent feedback loop (Burhan & Moradzadeh, 2020).

Secondly, the bias toward (mis)information that fits the position of the most engaged people contributes to the spread of such misinformation (Wittenberg & Berensky, 2020).

Thirdly, virality speeds up the diffusion of misinformation that is accepted uncritically by the users that are predisposed to believe certain contents, and so they quickly forward the messages.

Fourthly, in the case of targeted social media campaigns seeking to impact opinions, the massive use of programmed (ro)bots allows the distribution of the message to be amplified exponentially and reach a vast audience that could easily be overwhelmed by exposure to a myriad of messages tailored to the characteristics of the receivers.

Under such conditions, any attempt to counter the misinformation by posting reminders of basic facts is disadvantaged by the sheer size of the mass of misinformation. Subjectivity cancels objectivity. The massive use of bots in the forwarding of the messages is widely considered to have been a critical factor in the electoral victories of Trump in the US

(2016), Bolsonaro in Brazil (2018), the Brexit campaign in the UK (2016), Meloni in Italy (2022), as well as in the growing influence of such extreme right groups as Front National in France, Vox in Spain, and the Swedish Democrats.

It needs to be emphasized that these campaigns are very expensive and require resources and sophistication that only major business groups or powerful governments have at their disposal. Such was the case with the support of the American Koch brothers for Bolsonaro and of the Russian Government for Trump, and probably also for Le Pen and Meloni. In other words, this is not a blind automated process enacted by social media platforms, but a political strategy that skillfully uses the potential of social media and the vulnerability of the active users. The more people are engaged and the more they are susceptible to messages that they "like," then the more social media is influential.

The media platforms have an incentive to stimulate a polarized debate since it increases the traffic. The more users become active in the defense of their ideas and perceptions, and the more they are mobilized, positively or negatively, in the social media networks, the more data that can be captured or created.

When companies yield to the pressures of governments or of public opinion, they moderate some of the most extreme expressions of uncivility and misinformation through removal or a downranking of visibility in the algorithmically curated feeds. Complete removal through content takedown is extremely limited. For instance, in 2022, TikTok announced it removed over 110 million videos during April to June 2022, which accounts for less than 1% of the total number of videos posted during that period (TikTok, 2022). We should always keep in mind that the business model of most companies managing the digital platforms is based on increasing traffic in the networks, any kind of traffic. For this reason, they tend to prefer relying on curation that manages what content is most visible and for whom rather than full removal.[3]

---

[3]   For example, in 2018 Zuckerberg released a blueprint on content moderation where he acknowledges the conflicting incentives Facebook is subjected to: "One of the biggest issues social networks face is that, when left unchecked, people will engage disproportionately with more sensationalist and provocative content. This is not a new phenomenon. It is widespread on

And so the loop is closed: ideological and political polarization in society is concentrated in the activity of vocal minorities that hold on to their beliefs in dedicated spaces and relentlessly fight opposing views. In so doing, they contribute to further polarization in society, while enhancing the power and gains of the platforms by increasing user activity and providing a wealth of data that are appropriated and marketed by the owners of the networks. The technology masters of the world link up with oligopolistic business groups to establish profitable alliances with the political leaders of society, very often regardless of their ideological positions. There are, however, new instances of resistance to alternative projects and values enacted by social actors that are increasingly active in the public debate. Yet, by doing so, they increase further interaction and traffic, helping the accumulation of data for capitalism. At the same time, they broaden and intensify public debate, which often includes misinformation, that may undermine the political legitimacy of the power holders across the political spectrum. The political actors that are winners and losers in the debate may vary, but over time the legitimacy of the political system as a whole is eroded.

In sum: social media do not create political and ideological polarization, which has its roots in the conflicts within society. But they exacerbate and broaden polarization, using the digital public sphere as an amplifier of uncivil ideological confrontation.

References

Achiume, T. (2018) "Contemporary forms of racism, racial discrimination, xenophobia and related intolerance," A/73/305. United Nations. Available at: https://documents-dds-ny.un.org/doc/UNDOC/GEN/N18/251/48/PDF/N1825148.pdf?OpenElement (accessed on 4 February 2023).

---

cable news today and has been a staple of tabloids for more than a century. [...] Our research suggests that no matter where we draw the lines for what is allowed, as a piece of content gets close to that line, people will engage with it more on average—even when they tell us afterwards they don't like the content. [...] This is a basic incentive problem that we can address by penalizing borderline content so it gets less distribution and engagement" (2018).

Arsenault, A.H., & Castells, M. (2008) "The structure and dynamics of global multi-media business networks," *International Journal of Communication*, 2, pp. 707–748.

Bakshy, E., Messing, S., & Adamic, L.A. (2015) "Exposure to ideological diverse news and opinion on Facebook," *Science*, 348 (6239), pp. 1130–1132.

Barbera, P. (2020) "Social media, echo chambers, and political polarization," in N. Persily & J.A. Tucker (eds), *Social Media and Democracy: The State of the Field and Prospect for Reform*. Cambridge: Cambridge University Press, pp. 34–55.

Boxell, L., Gentzkow, M., & Shapiro, J.M. (2017) "Greater Internet use is not associated with faster growth of political polarization among US demographic groups," *Proceedings of the National Academy of Sciences*, 114(40), pp. 10612–10617.

Burhan, R. & Moradzadeh, J. (2020) "Neurotransmitter dopamine (DA) and its role in the development of social media addiction," *Journal of Neurology & Neurophysiology*, 11(7), pp. 01–02.

Castells, M. (2009) *Communication Power*. Oxford: Oxford University Press.

Castells, M. (2015) *Networks of Outrage and Hope: Social Movements in the Internet Age*. Cambridge: Polity.

Castells, M. (2017) *Rupture: The Crisis of Liberal Democracy*. Cambridge: Polity.

Edelman (2022) *2022 Edelman Trust Barometer*. Available at: https://www.edelman.com/trust/2022-trust-barometer (accessed on 27 April 2023).

Khan, I. (2021) "Report of the Special Rapporteur on the promotion and protection of the right to freedom of opinion and expression," A/76/258. United Nations. Available at: https://documents-dds-ny.un.org/doc/UNDOC/GEN/N21/212/16/PDF/N2121216.pdf?OpenElement (accessed on 4 February 2023).

Lakoff, G. (2008) *The Political Mind: Why You Can't Understand 21st Century Politics with an 18th Century Brain*. New York: Viking.

Nielsen, R.K., & Fletcher, R. (2020) "Democratic creative destruction? The effect of a changing media landscape on democracy," in N. Persily & J.A. Tucker (eds), *Social Media and Democracy: The State of the Field and Prospect for Reform*. Cambridge: Cambridge University Press, pp. 139–162.

OHCHR (2019) "Joint open letter on concerns about the global increase in hate speech." Available at: https://www.ohchr.org/en/statements-and-speeches/2019/09/joint-open-letter-concerns-about-global-increase-hate-speech (accessed on 4 February 2023).

Persily, N., & Tucker, J.A. (eds) (2020) *Social Media and Democracy: The State of the Field and Prospect for Reform*. Cambridge: Cambridge University Press.

Sardinha, L., Maheu-Giroux, M., Stöckl, H., Meyer, S.R., & García-Moreno, C. (2022) "Global, regional, and national prevalence estimates of physical or sexual, or both, intimate partner violence against women in 2018," *The Lancet*, 399(10327), pp. 803–813. https://doi.org/10.1016/S0140-6736(21)02664-7.

Taplin, J. (2023) *The End of Reality: How Four Billionaires Are Selling a Fantasy Future of the Metaverse, Mars, and Crypto*. New York: PublicAffairs.

TikTok (2022) *Community Guidelines Enforcement Report*. Available at: https://www.tiktok.com/transparency/en-us/community-guidelines-enforcement-2022-2/ (accessed on 11 October 2022).

Touraine, A. (2012) *Pouvons-nous vivre ensemble?* Paris: Seuil.

Tucker, J.A., Guess, A., Barbera, P., Vaccari, C., Siegel, A., Sanovich, S., Stukal, D., & Nyhan, B. (2018) "Social media, political polarization, and political disinformation: A review of the scientific literature." Available at SSRN: http://dx.doi .org/10.2139/ssrn.3144139.

Wittenberg, C., & Berensky, A.J. (2020) "Misinformation and its correction" in N. Persily & J.A. Tucker (eds), *Social Media and Democracy: The State of the Field and Prospect for Reform*. Cambridge: Cambridge University Press, pp. 163–198.

World Health Organization (2021) *Violence against Women Prevalence Estimates, 2018*. Available at: https://www.who.int/publications/i/item/9789240026681.

Zuckerberg, M. (2018) "A blueprint for content governance and enforcement," *Facebook*. Available at: https://www.facebook.com/notes/mark-zuckerberg/a -blueprint-for-content-governance-and-enforcement/10156443129621634/ (accessed on 15 December 2020).

10 War and peace in the time of digital machines

War and technology have always been closely interrelated throughout history. Every major war, cold or hot, most significantly World War II, has introduced a wave of technological innovations that sometimes diffused into the entire realm of human activities. Yet, at the heart of all societies, power relationships are the determining factor. Power is ultimately institutionalized in the State. And the State, following the widely accepted concept proposed by Max Weber, is defined by its monopoly of violence. I would add: legitimate or not. Thus, regardless of the claims about the goodness of technology for humankind, a less kind humanity has always privileged the potential military applications in the advancement of technology. With the acceleration of innovation in digital technologies and the pervasiveness of computerized communication networking, a new form of warfare has emerged in the twenty-first century, and is already practiced in Ukraine at the time of writing. More, much more looms on the horizon. Other technologies of war-making are still current, be it massacres by machete or the threat of nuclear or biological extermination. However, all forms of mutual destruction are now incorporated in a digital environment that profoundly modifies and amplifies the implementation of our collective death wish.

There are different applications under the generic term "digital war," such as automated war-making operations, semiautonomous surveillance and killing machines, military planning facilitated by artificial intelligence, missiles and precision munition, electronic warfare (disabling communications), cyberwar (disrupting the digital and physical infrastructure on which societies are based), and disinformation and manipulation of public perception.

But all modalities of this kind of war are based on a vast system of digital communication networks that is expanding at an exponential rate. Without this networking capability between command-and-control centers and operatives on the ground, none of these sophisticated weaponry and disruption mechanisms could function.

Digital communication networks

The skies of our planet are now populated by thousands of satellites, both in low orbits and in outer space, that connect every activity on Earth. The depths of our oceans harbor thousands of kilometers of fiber-optic cables that keep the beat of the global flow of information upon which we rely. Together they provide the fundamental infrastructure of our societies, both for peace and for war.

In 2022, about 6,000 human-made objects were orbiting the planet. Their number increases exponentially. Governments used to be the agencies in charge of launching and operating satellites; however, most of the industry is now being privatized. Of the 6,000 objects, 4,047 were for commercial use, while the military accounted for 424 and government 520. The US is by far the leader in the ownership of satellites, with 3,145, while China controls 535, Russia 170, and multinational consortia 180 (Statista, 2023). The privatization of the satellite launching and operating activity is exemplified by Elon Musk's SpaceX, which in 2022 owned 1,919 satellites, and in 2022–23 added Internet satellites, so increasing in number by 185%, a rate that has most likely accelerated since the Ukraine war. Many of the Starlink satellites (the SpaceX system) have been contracted for service by NASA and the US Government. Japan has also engaged the company to launch a similar collaboration (Reuters, 2023). The SpaceX Starlink system provided support to the Ukrainian troops on the ground, as well as maintaining the infrastructure of the country under Russian bombardments. Although in the early stages of the war Musk tried to refrain from open hostility toward Russia, refusing to provide service for drone attacks (BBC, 2023), later on the US Government appears to have contracted Starlink's services to guide the Ukrainian drones on the front lines and in attacks on Russian territory. The strategic importance of satellite-based communication for surveillance, guidance, and coordination in military operations obviously opens the possibility for attacks on the satellite system. Some attacks could be direct hits by missiles or collisions, others by falling debris from a destroyed satellite (Shepherd, 2022). Satellites are also submitted to electronic warfare by jamming communication or corrupting their software. Starlink has accused Russia of conducting such operations in the context of the Ukraine war.

While satellite communication appears at the forefront of global communication networks, in fact undersea cables are much more important in this regard. Ninety-five percent of traffic on the global Internet is carried by about 200 undersea fiber-optic cables. Both governments and companies own and operate these cables. Because of their critical role in keeping everything working in a globally networked economy, their vulnerability has become a major security concern (Ratiu, 2021). This is because, in addition to the possible physical damage they can suffer, they can also be tapped for spying or disinformation. Thus, governments have engaged in developing a new form of submarine warfare to attack and protect the cable system (Garret, 2018; Long, 2023). The USS *Jimmy Carter* has been operating since 2005, acting as a mothership for unmanned submarine vehicles to perform diverse tasks. It will be followed by the most expensive nuclear submarine ever built, costing $5 billion, which is already under construction in Groton, Connecticut, with a wide array of capabilities that include offensive action as well as the preservation of the cable network (Meyer, 2023). Russia has responded to these challenges by introducing in July 2022 the *Belgorod*, the longest nuclear submarine in the world, fully equipped with electronic warfare equipment, as well as armed with two megaton nuclear torpedoes that can be launched from long distances without being detected. China is following the same track.

On the basis of this dense system of global-local communication networks, new technologies of war-making have been introduced into the military operations currently underway and projected for the near future.

Drones

Drones became the weapons of choice in the Middle Eastern wars and in the Ukraine war, both for reconnaissance tasks and for strikes. Originally, in the late 1960s, Israel introduced drones as an instrument of surveillance and electronic warfare, becoming decisive in the air battles with Syrian MiGs in 1982. The US soon adopted the Israel-made Pioneer unmanned aerial vehicle (UAV), and generalized the use of drones after 9/11 to locate targets and conduct assassinations in the Middle East and Afghanistan. The US deployed the Predator, armed with Hellfire missiles. The diffusion of drone technologies has already reached most countries and for a variety of uses, particularly in agriculture, transportation, and disaster relief. Turkey and Iran have developed large export markets. As did China thereafter. The US developed new models under the generic name of MALE (Middle Altitude Long Endurance), supplying France among other countries. In the Ukraine war, highly sophisticated drones have been introduced by the warring parties. Ukraine successfully used the Turkish-made Baryaktar TB2 to destroy tank columns at the onset of the Russian invasion. Russia countered by focusing on low altitude air defenses and electronic jamming, ultimately forcing the Ukrainians to withdraw this model. Similarly, Russia introduced two very advanced drones in the first months of the war, the Forpost and the Orion, that became vulnerable to the Ukrainian air defenses and were also side-lined later on. Their vulnerability is linked to their need to fly low to be effective and to the insufficient defense against electronic warfare. Thus, both armies adopted new tactics, using large numbers of simple-purpose drones, monikers, or suicide drones used as flying bombs (DeVore, 2023). Indeed, cost has been a major consideration for the wide use of drones, together with the effort to save pilots' lives. The use of cheap drones meant that it often became more expensive to shoot one down than the cost of the drone itself. Ukraine relied on the reconnaissance capability of drones to support the tactical advance of small fighting units across the dense Russian defense systems. It was also able to attempt long-distance drone attacks in Russian territory, including Moscow. Russia used drones in large groups, trying to overwhelm Ukrainian air defenses, often performing missile attacks under the cover of the drones. Russia has been using, and losing, hundreds of a relatively simple Iranian drone, the Shahed (Witness)-136, now being mass produced in Russia under a license from Iran.

A more advanced form of drone attack is swarming, which implies not only large groups of drones, but their capability to communicate among themselves and make semiautonomous decisions based on AI technology (Jankowicz, 2023). This technology exists already in the experimental stage, as developed by Raytheon for the US Department of Defense. A new generation of air drones is being developed by the US Navy in cooperation with the leading aviation/missile companies, including the X-47B UCAS, the Sea Ghost, the Phantom Ray, and the Predator C Avenger.

Drones are not limited to air operations. In fact, their fastest development is in naval and submarine warfare. Ukrainians, advised by British experts, and with the support of NATO satellite surveillance, have successfully used marine suicide drones against Russia warships and port infrastructure in Crimea.

A number of misrepresentations have surrounded the generalization of drone warfare. One of them is that operations are fully conducted by remote control. In fact, to be effective, drone operators have to be connected to ground personnel on the front line or in nearby air bases (Jeangène Vilmer, 2023). For instance, for a 24-hour shift operating the American MQ-9 Reaper, 250 personnel are required, of which 61 are forward deployed in the area of operation. Another, and more sensitive, debate concerns the faceless killings performed with this new weapon. True, many of the drone assassinations have been conducted at distance from air force bases in the Nevada Desert. However, those operating the drones and sending in the missiles have a very clear vision of what they are doing, which sometimes requires recognizing the face of the target. To be sure, collateral damage, in terms of killing innocent people, often results from the attack. But this destruction is also witnessed vividly by the perpetrators. In fact, fighter pilots exchanging missile fire hundreds of miles from their targets have a more abstract experience of war than the very real virtuality of drone operators. Much of the negative public opinion against the use of drones derives from their use in extraterritorial clandestine attacks, which are taking place without any kind of regulation or judicial oversight.

The vision of an automated, robotic war has materialized. But it is not taking the shape of humanoids marching in battalions. It is being conducted by flying or swimming objects that search and destroy on command in real time.

AI warriors

The extraordinary improvements in artificial intelligence, as noted in the Introduction of this volume, have deeply impacted warfare as they have all other domains of human life. Nonetheless, to put matters into perspective we should consider that artificial intelligence operates on the basis of available data. While the recombining and analytical capacity of AI programs improves the capability of processing information and modeling complex systems in an operational form that can be useful for decision making, AI is only as good as the data that it is exposed to, both in training and in action. So, network connectivity and computing power will determine its actual impact in real life, as in real death, when applied to war, ultimately a death machine. In fact, the influence of AI in military practice varies depending on the realm of its application.

A major focus concerns the pursuit of autonomous or semiautonomous reconnaissance and fighting machines. Drones are the most direct manifestation of this goal. Stimulated by the Ukraine war, Russia and the US are investing major resources in designing and operating semiautonomous machines that can be air-, water-, or land-based. Russia announced that its Lancet technology is operational, and so did the US, with its own program, named Switchblade. Yet, these machines will be working under human control for the foreseeable future.

More powerful AI programs will perfect the embedded navigation systems in ammunition (dubbed "smart bombs") that have been used since the Gulf war and filtered down to the wide diffusion of GPS navigation systems that are used daily by civilians. Powerful, mobile artillery is having a new life in the Ukraine war, thanks to the capacity to target the enemy effectively, by relying on timely drone and satellite information combined with high-speed decision-making assistance from AI. The US Himars long-distance rockets have overwhelmed the traditional mass bombardment tactics of the Russian Grads multi-rocket systems inherited from World War II. Long-distance, high-precision missiles have shifted much of the firepower to exchange of missiles and anti-missile defense. A generation of cruise missiles, and particularly Russian hypersonic missiles, are able to search for their targets, select their routes, and evade defenses. However, the high cost of the weapon has limited its use by Russia and has kept it mostly in reserve by NATO for the time being. The capability to fire at distance with accuracy is transforming the role of helicopters in combat. The defeat of the Soviet Union in Afghanistan was facilitated by the introduction of American Stinger portable missiles that, in the hands of the Mujahidin, decimated the fearsome Soviet flying machines when they intervened in close fighting. Something similar took place in the early stages of the Ukraine war, when the Stingers again repelled the attacks of Russian helicopters. One year into the war, Russia introduced a new model of attack helicopter, the Alligator, capable of firing powerful missiles with great precision at a distance of several miles, out of the range of Stingers.

Yet, perhaps the most significant changes linked to military AI are related to analyzing at high speed in real time options of military tactics, be it at the level of the battlefield or on specific small-scale operations. The decentralization of data and decision-making capability is allowing the development of a network model of armed forces, in which small, agile units with considerable firepower, coordinate their action and evolve depending on changing circumstances. Vertical command-and-control, characteristic of the Russian tradition, is being seriously challenged by the versatility and autonomy of networked military forces, able to combine in real time multiple forms of action, including reconnaissance, and air and artillery support (Leitzel & Hillebrand, 2022).

Strategic military planning is hugely benefitting from the different options offered by AI models to guide the course of action on different timescales. War games become practical tools when provided with the flexibility to adapt to the flow of data coming from experiences on the ground. However, good decision making ultimately depends on the accuracy of the information on which AI operates. Here there is a major transformation in the information-gathering process in and around the battlefield: what has been labeled as "participatory information warfare."

Participatory information, war, and digital networks

People are now capable of appropriating for their practices a variety of communication networks. This empowerment allows them to participate in the retrieval and distribution of information on the ongoing fighting. For those closest to the front line, this includes relaying information of what they can directly observe to the fighting units they support, as has often been the case in Ukraine. It is a form of mass spying in and around the battlefields, with a lesser risk of being detected at the time of action. It certainly depends on the empathy between the population and the armed forces. Thus, it gives the edge to the invaded over the invaders. This is a most traditional form of resistance to superior force that has always been essential in partisan and guerrilla warfare. Now it can be multiplied and sharpened by the widespread use of social networks. The massive use of Telegram for this purpose during the Ukraine war exemplifies this significant development of the connection between war and society.

Participatory information warfare can also be enacted at distance. In the Ukraine war, there was a flurry of both Russian and Ukrainian military bloggers who were often the only sources of information on what was actually happening in the war. Indeed, both governments imposed a news blackout that made it extremely difficult for traditional mass media to inform about the reality of war. Embedded journalism became the only way to report, and propaganda often took over the content of the broadcast. While bloggers were also partisan in most cases, they also tried to report what was happening, particularly in the Russian case, to serve the public in such a crucial situation. Obviously, information and disinformation are mixed in varying proportions depending on sources and themes. However, both forms of intervention became important factors in the formation of public opinion, ultimately conditioning the conduct of the war.

Cyberwar

Cyberwar refers to intrusion in or disruption of computer networks and computers themselves. It is often conducted through malware, that is software that destroys or corrupts files and programs. It can be silent, for the purpose of spying, or it can involve a full-fledged attack that disables the networks that run essential activities in every domain. Although the military is mainly concerned about hacking into their computer systems, a greater concern is the overall cyber-infrastructure on which everything depends in our digital societies, including electricity, water, transportation, health, hospitals, schools, government, banks, business, entertainment, and science. Attacks can be performed by hostile governments, but also by anonymous hacktivists of diverse ideologies and different purposes, as well as common criminals focused on stealing or locking data and demanding ransom from institutions and companies. It is in fact a pervasive practice that has led to a constant deployment of electronic defenses in all networks. Yet, some of the most powerful information systems, be it the Defense Department or Microsoft, have seen their defenses penetrated. The reason is very simple. Protection is as vulnerable as the weakest point in the network. Thus, both companies and the government have been hacked by breaking the electronic defense of a contractor or employee connecting at distance with the core computers in order to do their work. Firewalls and encrypted access keys have become increasingly sophisticated and redundant. No stable solution has been found. The better the defenses, the better the technologies of disruption and the talent of the disruptors. In fact, it is surprising that no major catastrophes have been recorded to date given the breadth of critical infrastructure vulnerable to attacks and the many choke points. It is a matter of time until airports will crash at a high traffic moment, or blackouts of entire regions will recur with greater frequency than in the accidental events of the past. For those geopolitical actors engaging in asymmetrical confrontation, cyber terror—targeting the networks sustaining our lives—provides a major opportunity. Some catastrophes have already been averted at the last minute, although there is a necessary veil of discretion on the matter. We must be aware that one of the most dramatic hidden forms of warfare in our time is being fought daily between security agencies, digital safety experts, and crackers and counter-crackers (note: hackers are not crackers). Most countries actively participate in this form of conflict, with different goals and effectiveness, as shown in Tables 10.1 and 10.2, which show information from Harvard's Belfer Center (2020).

Table 10.1 National Cyberpower Index, Top 10

Position	Country	Overall score	Capability ranking	Intent ranking
1	United States	50.24	1	2
2	China	41.47	2	1
3	United Kingdom	35.57	3	3
4	Russia	28.38	10	4
5	Netherlands	24.18	9	5
6	France	23.43	5	11
7	Germany	22.42	4	12
8	Canada	21.5	11	9
9	Japan	21.03	8	14
10	Australia	20.03	16	8

Note: The overall NCPI assessment measures the "comprehensiveness" of a country as a cyber actor. Comprehensiveness, in the context of NCPI, refers to a country's use of cyber to achieve multiple objectives as opposed to a few. The most comprehensive cyber power is the country that has (1) the intent to pursue multiple national objectives using cyber means and (2) the capabilities to achieve those objective(s).
Source: Designed by Juan Ortiz Freuler based on information from the Harvard Belger National Cyber Power Index, published in 2020 by the China Cyber-Policy Initiative.

Table 10.2 Ranking of cyber-intent, by objective

	Surveillance	Defense	Control	Intelligence	Offense	Norms
1	Russia	UK	US	UK	UK	UK
2	China	Netherlands	China	US	US	Germany
3	Vietnam	France	Russia	Spain	Israel	US
4	Saudi Arabia	US	Vietnam	Netherlands	Spain	Japan
5	UK	China	Israel	Israel	Russia	France
6	Estonia	Japan	Iran	Russia	Iran	Switzerland
7	Netherlands	Canada	UK	New Zealand	China	Netherlands
8	Australia	Sweden	Germany	Canada	Netherlands	China
9	US	Estonia	New Zealand	Australia	Estonia	Canada
10	Switzerland	Australia	France	China	Australia	Australia
		Russia (#17)				*Russia (#14)*

Note: The objectives are surveilling and monitoring domestic groups; strengthening and enhancing national cyber-defenses; controlling and manipulating the information environment; foreign intelligence collection for national cybersecurity; commercial gain or enhancing domestic industry growth; destroying an adversary's infrastructure and capabilities; and defining international cyber-norms and technical standards.
Source: Designed by Juan Ortiz Freuler based on information from the Harvard Belger National Cyber Power Index, published in 2020 by the China Cyber-Policy Initiative.

Quantum war

Digital war in its different manifestations ultimately depends on computing power. Throughout this volume, I have underlined the significance of quantum computing for a complete transformation of the process of digitization, albeit mature applications are forecasted only for the 2030–40 timeframe. Military strategists are investing heavily in accelerating the applications of quantum computing to warfare and strategic planning (Krelina, 2021; USNI, 2021), particularly China. According to the US Defense Department, the following appear to be the most direct applications of quantum computing for military activities (van Amerongen, 2021).

The first is the development of quantum sensors capable of detecting objects invisible to radar until now, such as nuclear submarines and stealth airplanes. This capability extends to the development of "Position, Navigation, Timing" (PNT) that allows navigation without relying on external references such as GPS.

The second breakthrough eventually made possible by quantum computing concerns the ability to decipher highly sophisticated encryption, such as the widely used RSA algorithm, through efficient factorization and solving the complex mathematical equations that underlie state-of-the-art encryption that protects our online communication. In fact, there is already an algorithm able to perform this task: the Shor's algorithm, discovered in 1994, can efficiently factorize large numbers. However, the algorithm has been waiting for the advent of quantum computing, as calculations of the computational complexity involved have remained a challenge until now.

The third major application is the design of secure networking for the transfer of data through "Quantum Key Distribution" (QKD) technology. China has targeted this technology as a priority, and has already laid out an experimental network based on this principle.

Furthermore, quantum computing can greatly enhance any AI application based on computation capacity, therefore improving AI applications in the whole realm of warfare needs and strategies. Particularly important in this regard is the improvement of machine learning, and the acceleration of autonomous war-making machines and information systems. Indeed, any major technological innovation is bound to affect and interact with the entire landscape of digital warfare.

Change and continuity in warfare

The reality of war mixes all the forms of digital war outlined in this chapter. It is what some experts label "hybrid war." But the actual conduct of violent conflict includes the modalities of another time, low tech and high tech. The Ukraine war has seen the return of war in the trenches along hundreds of miles of front line, with thousands of casualties involved in moving the line just a few meters. World War I and digital war merge in the daily routine of killing. The battle for Bakhmut made no strategic sense for either of the fighting factions. The declared purpose on both sides was to weaken the enemy by inflicting the largest losses possible, alongside the traditional macho bravado of planting a flag on the ruins of a devastated city.

Precision bombardments, particularly those performed by Russia, did not spare civilian lives, which were lost by the thousands, be it as collateral damage or deliberate terror strategy. In fact, this war, which cannot be fully won by either side, served as a testing ground for the new digital weapons and tactics of our age, both for Russia and for NATO, while keeping in the background the potential use of tactical nuclear weapons, and the possibility of an out-of-control escalation. War, in all ages, has always been death, destruction, and horror. The more technology advances, the more destructive and cruel war becomes.

References

BBC (2023) "Ukraine war: Elon Musk's SpaceX firm bars Kyiv from using Starlink tech for drone control," *BBC News*, 9 February. Available at: https://www.bbc.com/news/world-europe-64579267 (accessed on 30 June 2023).

Belfer Center (2020) *National Cyber Power Index 2020*. Harvard Kennedy School of Government. Available at: https://www.belfercenter.org/publication/national-cyber-power-index-2020 (accessed on 25 October 2023).

DeVore, M.R. (2023) "'No end of a lesson:' Observations from the first high-intensity drone war," *Defense & Security Analysis*, 39(2), pp. 263–266. https://doi.org/10.1080/14751798.2023.2178571.

Garret, H. (2018) "Evaluating the Russian threat to undersea cables," *Default*, 5 March. Available at: https://www.lawfaremedia.org/article/evaluating-russian-threat-undersea-cables (accessed on 30 June 2023).

Jankowicz, M. (2023) "Military tech is racing towards a dangerous AI future, and Russia's war in Ukraine is paving the way, drone experts say," *Business Insider*. Available at: https://www.businessinsider.com/drone-ai-russia-ukraine-war-experts-concerned-for-future-2023-1 (accessed on 29 June 2023).

Jeangène Vilmer, J.-B. (2023) "Not so remote drone warfare," *International Politics*, 60(4), pp. 897–918. https://doi.org/10.1057/s41311-021-00338-9.

Krelina, M. (2021) "Quantum technology for military applications," *EPJ Quantum Technology*, 8(1), p. 24. https://doi.org/10.1140/epjqt/s40507-021-00113-y.

Leitzel, B.C., & Hillebrand, G. (2022) *Strategic Cyberspace Operations Guide*. Carlisle, PA: US Army, Center for Strategic Leadership. Available at: https://csl.armywarcollege.edu/USACSL/Publications/Strategic_Cyberspace_Operations_Guide.pdf.

Long, M.L. (2023) "Information warfare in the depths: An analysis of global undersea cable networks," *U.S. Naval Institute Proceedings*, 149(5), 1443. https://www.usni.org/magazines/proceedings/2023/may/information-warfare -depths-analysis-global-undersea-cable-networks (accessed on 30 June 2023).

Meyer, J. (2023) "Under the sea and ready for war? US wants to spend billions on spy submarine to fend off ocean-deep China, Russia advances," *USA Today*, 23 May. Available at: https://www.usatoday.com/in-depth/news/politics/2023/ 05/23/seabed-warfare-new-combat-us-russia-china/70208034007/ (accessed on 30 June 2023).

Ratiu, A. (2021) "Cyber defense across the ocean floor: The geopolitics of subma-rine cable security," *Atlantic Council*, 13 September. Available at: https://www .atlanticcouncil.org/in-depth-research-reports/report/cyber-defense-across -the-ocean-floor-the-geopolitics-of-submarine-cable-security/ (accessed on 30 June 2023).

Reuters (2023) "Japan's military considers adopting Musk's Starlink satellite service, Yomiuri newspaper reports," *Reuters*, 25 June. Available at: https:// www.reuters.com/technology/japans-military-considers-adopting-musks -starlink-satellite-service-media-2023-06-25/ (accessed on 30 June 2023).

Shepherd, T. (2022) "'Life no longer as we know it': War in space would have dev-astating effects, military experts say," *The Guardian*, 28 November. Available at: https://www.theguardian.com/science/2022/nov/29/life-no-longer-as-we -know-it-war-in-space-would-have-immediate-effects-expert-says (accessed on 29 June 2023).

Statista (2023) "Infographic: The countries with the most satellites in space," *Statista Infographics*. Available at: https://www.statista.com/chart/17107/ countries-with-the-most-satellites-in-space (accessed on 29 June 2023).

Tucker, J.A., Guess, A., Barbera, P., Vaccari, C., Siegel, A., Sanovich, S., Stukal, D., & Nyhan, B. (2018) "Social media, political polarization, and political disinfor-mation: A review of the scientific literature." Available at SSRN: http://dx.doi .org/10.2139/ssrn.3144139.

USNI (2021) "Report on military applications for quantum computing," *USNI News*, 27 May. Available at: https://news.usni.org/2021/05/27/report-on -military-applications-for-quantum-computing (accessed on 30 June 2023).

van Amerongen, M. (2021) "Quantum technologies in defence & security," *NATO Review*, 3 June. Available at: https://www.nato.int/docu/review/articles/2021/ 06/03/quantum-technologies-in-defence-security/index.html (accessed on 30 June 2023).

11 Conclusion: the digital society and the network society

The digital society is the socio-technological form that underlies the coming of age of the network society while, in return, being shaped by the dynamics of the network society.

In the twenty-first century, a new social structure has emerged worldwide as the prevalent form of organization in all domains of human life. I conceptualized this social structure, some time ago, as the network society because its defining feature is that all core activities are organized in networks, whose logic permeates the entire realm of human practices. Although networks have always existed, the strategically important contemporary networks that articulate the social structure are powered by digital communication technologies, which increase exponentially the flexibility, scalability, and resilience of these networks. At the source of the network society, there are several economic, social, cultural, and institutional processes that interacted and ultimately converged in the genesis of this new form of human existence. A global economy, whose sources of productivity are primarily dependent on knowledge, information, and communication. A culture of individuation that emphasizes autonomy of the subjects. Networked social movements oriented toward a transformation of social values in opposition to hierarchical forms of social organization. The challenge to patriarchal gender relations. The holistic view of human life as interdependent with nature, giving priority to the preservation of a planet that can sustain life, in direct contradiction to the one-sided productivist logic of the industrial age, thus asserting inter-generational solidarity. New time/space configurations as support of social practices: space of flows and timeless time (Castells, 2000 [1996]; 2004). These transformations relied on the networking form of organization as an appropriate foundation for their materialization. The deployment of the multidimensional network structure was made possible by the socio-technological platforms that resulted from the revolution in ICTs.

I conceptualized the new technological paradigm originated by this revolution as informationalism, because it was characterized by new capacities in processing and distributing information, everywhere and in everything, in similar ways to what had been defined as industrialism, resulting from the technological revolution in the capacity to generate and diffuse energy (Touraine, 1969; Bell, 1973). Neither the sociocultural factors nor the technological transformations were the producers of the network society. The genesis of the new social structure came from the interaction of both transformative processes. Their coincidence was the result of differential histories that happened to converge in a given time. However, as in the case of the industrial society, it took some time before digital technologies would evolve and diffuse to reach a critical point that supported the full deployment of the networking form of social organization. Nowadays our society is characterized by a digital DNA (Cowhey & Aronson, 2017).

The digital society: dawn or doom?

The network society and its socio-technological platform, the digital society, process values and institutions that are the result of human action. The prevalence of certain values is a function of power relationships in each society. As industrial societies came in different institutional forms, such as capitalism(s) or statism(s), with diverse effects on human experience, the network society manifests itself in distinct social organizations, according to the history and culture of each country and to its specific position within the global networks that organize life on the planet. Similarly, the digital society, in its wide array of new technologies, has different, and even opposite, effects on humans and nature, depending on who uses each technology and for what it is used. For instance, the exponential growth of artificial intelligence, enabling machines to identify emergent properties for analysis, communication, and decision making, has been shown to result in diverging outcomes depending on the social practices that embody the technology (Crawford, 2021; Ferrara, 2024). There is widespread concern among the public concerning the fast diffusion of powerful technologies that can change our existence without knowing the actual effects of the forces we have unleashed. Thus, the urgent calls coming from many quarters, including from those who invented and provided the technology, for a moratorium on their use,

until we have enough knowledge and policy guidelines to govern technological transformation, before the promise of a new dawn turns into the
unforeseen doom of our species. The intellectual project underlying this
book is an attempt to move from disinformed bewilderment to informed
consciousness of our new human experience.

References

Bell, D. (1973) *The Coming of the Postindustrial Society*. Cambridge, MA: Harvard
University Press.
Castells, M. (2000 [1996]) *The Rise of the Network Society*, 2nd edn. Oxford:
Blackwell.
Castells, M. (2004) "Informationalism, networks, and the network society: A the-
oretical blueprint," in M. Castells (ed.), *The Network Society: A Cross-Cultural
Perspective*. Cheltenham, UK and Northampton, MA, USA: Edward Elgar
Publishing, pp. 3–45.
Cowhey, P., & Aronson, J. (2017) *Digital DNA*. New York: Oxford University
Press.
Crawford, K. (2021) *Atlas of AI*. New Haven, CT: Yale University Press.
Ferrara, E. (forthcoming in 2024) "GenAI against humanity: Nefarious applica-
tions of generative artificial intelligence and large language models," ACM.
Touraine, A. (1969) *La société post-industrielle*. Paris: Denoel.

Index

5G 6, 95

advertising 36–7
age, and digital divide 97–8, 105
Alphabet 36
Amazon 17, 18, 19, 20, 37
American Civil Liberties Union
    (ACLU) 40
Android 18
Ant 61–2
Apple 18, 19, 20, 37, 44, 96
Arab Spring 111, 119
artificial intelligence (AI) 5, 19–20, 23
    drones 135
    and employment 6–7
    and fake news 25
    financial markets 54
    learning and education 85–9
    self-driving cars 43
    translation 104
    war 135, 136–8, 142
Assange, Julian 39
autonomy 14–15, 114–15

Bezos, Jeff 45–6
'big tech' 20
bin Salman, Mohammed 46
Bitcoin 56–9, 61, 62
blockchain technology 57–8, 60–61
books (print) 21, 22
bots 127–8
Boxell, L. 126–7
boyd, d.m. 15
broadband 94, 101–2, 105
business models
    based on users' information 36–8

diversification 44
new 21–4

cables 133–4
Cambridge Analytica 39
capitalism 55
    data capitalism 37–8, 125
    informational capitalism 31, 32,
        37–8
Castells, M. 14–15, 54, 111, 113, 124
Central Bank digital currencies
    (CBDCs) 54–5, 62
centralization of information 31
ChatGPT 23, 25, 43, 86, 87–8, 104
Chaum, David 56
Citizen Lab 45
class, and digital divide 99–101
Claude GPT 86–7
cloud 3–4, 94
collaboration 19
communication 9–11
    decreasing monopoly over 31
    social dynamics in mass
        communication 122–3
'communication society' 11
communities of practice 117
connectivity 5–6, 94
Covid-19 pandemic 26, 67, 68, 73–5
credit default swaps (CDS) 52–4
crisis of legitimacy 122, 124
cross-collaboration 19
crypto culture 56
crypto-wars 39
cryptocurrencies 55–60, 61–2
cultural digital divides 102–4
currencies 54–62

cyber-intent 141
cyberwar 139–41
cypherpunks 56–7

data capitalism 37–8, 125
dating 13–14
'deep fake' 25
derivatives 52–4
digital currencies 54–62
digital divides 93
    age 97–8
    class 99–101
    cultural 102–4
    ethnicity 101–2
    gender 95–7
    meaningful access to the Internet
      105
    territorial 93–5
digital military technologies *see* war
digital nomads 71–3
digital social networks 15–16, 17
digital society and network society
    145–7
digital tokens 60–61
digital trading *see* financial markets,
    digitalization of
digitization 1
Dignity 116
discrimination 102
disinformation and misinformation
    25, 26–7, 83, 124–5, 127–8
Disney 19
distance learning 84–5
dopamine 127
drones 134–6, 137

e-books 22
'echo chambers' 126
economies of scale 66
education
    level of, and acceptance of fake
      news 26
    see also learning
educational background of family 80,
    99
Electronic Frontier Foundation (EFF)
    40
Ellison, N.B. 15

employment 6–7
encryption 39, 40, 41–2, 44–5, 142
English language, use of 102–4
Ethereum 60
ethnicity, and digital divide 101–2

Facebook 15–16, 17, 18, 20, 21–2, 44,
    96, 102, 128–9
fake identities 13
fake news 24–7, 83, 124–5
Fernández-Ardèvol, M. 98, 100
Financial Automaton 54
financial crisis 2008–10; 53, 54
financial markets, digitalization of
    blockchain platforms, digital
      currencies, and smart
      contracts 60–61
    cryptocurrencies 55–60
    securitization 52–5
    State and cryptocurrencies 61–2
Five Eyes cooperation program 32–3
Fourteen Eyes cooperation program
    33
Freuler, Ortiz 100, 101

Gaetz, Matt 42
gender gap 95–7
Giddens, A. 15
gold 55
Google 17, 18, 20, 41–2, 44, 96

Hansen, S. 68, 69
Hayden, Michael 33
helicopters 137
high frequency trading (HFT) 54
Hughes, Eric 57
Hulu 19
human rights 116

individuation 15
inequality 99–101, 111–12, 124
information
    collection, commodification, and
      processing of 36–8
    disinformation and
      misinformation 25, 26–7,
      83, 124–5, 127–8

informational capitalism 31, 32, 37–8
infrastructure 4, 94–5, 100
Instagram 17, 96–7
Internet
    advent of 9
    expansion of 1–2
    gender gap 95–6
    languages used on 16, 102–4
    meaningful access to 105
    penetration 1, 2, 93, 94
    and sociability 11–15
Internet of Things (IoT) 2–3
intra-metropolitan decentralization
    66, 71, 74
iOS 18

Koo, Nahoi 12–13

L3Harris 46
languages used on Internet 16, 102–4
Large Language Models (LLMs) 5, 20,
    86, 104
latency 6
learning 79
    academic performance, factors
        favoring 79–80
    artificial intelligence (AI) and
        machine learning 85–9
    computers in education 79–84
    distance learning 84–5
    use of smartphones 99
    virtual classroom 89
legitimacy, crisis of 122, 124
libertarians 39, 45, 55
life satisfaction 12–13
low orbit space 43

machine learning 86–9
market capitalization 19, 20
marriage 14
mass self-communication 9, 31
Meta 16, 19, 36, 44
metropolitan decentralization 66, 70,
    71, 74
metropolitan region 65–7, 70–71
Microsoft 19–20, 23, 37, 43
military bloggers 138–9

military tactics and planning 137–8
millennials 70
mining 58–9, 61
misinformation and disinformation
    25, 26–7, 83, 124–5, 127–8
missiles 137
mobile phones 1, 2, 99, 105
    see also smartphones
money 55, 61
multimedia business networks 16–21
Musk, Elon 22–3, 24, 43, 133

National Cyberpower Index 140
National Security Agency (NSA) (US)
    32, 33–5, 38–9, 40–41
negative news 25–6
network society and digital society
    145–7
'networked individualism' 14
networked social movements 110
    anatomy of 113–16
    communication technologies
        116–18
    early twenty-first century 111–13
    and political change 118–20
networks
    digital communication 133–4
    digital social 15–16, 17
    metropolitan regions 66–7
    multimedia business 16–21
newspapers 20–21
Nine Eyes cooperation program 32–3
NSO 45, 46

Occupy Wall Street movement
    111–12, 119
OECD studies 81–3, 96
Onion Router (TOR) 39–40
online dating 13–14
OpenAI 19–20, 23, 43
outmigration 70, 73–4

Palantir 42
participatory information warfare
    138–9
pay in digital industries, gender gap 96
'PayPal mafia' 22–3

Pegasus 45–6
plagiarism 88
political change, and networked social movements 118–20
political polarization 13, 122–9
power relationships 122, 132
PRISM program 34, 35
privacy *see* surveillance/State surveillance
Project Maven 42

quantum computing 6, 141–2

Randazzo, Marley 13–14
remote work *see* teleworking and remote work
Reset the Net 41
rhizomatic social movements 115
rumors 24–5

satellites 133
'Satoshi Nakamoto' 57
Saudi Arabian intelligence 46
Second Life 22
securitization 52–5
sex 14
Signal 40
smart contracts 60–61
smartphones 10–11, 45–6, 93, 94, 96, 97, 98, 99, 102, 114, 118
Snowden, Edward 34–5, 38
sociability 11–15
social autonomy 14–15
social media
    and discrimination 102
    and gender 96–7
    and political polarization 122–9
    time spent with per day 10
    use of term 10, 15
social movements *see* networked social movements
social networks, digital 15–16, 17
SpaceX 133
spatial concentration 65–6, 70–71
spatial structure, emerging 70–71
stable coins 61
State, and cryptocurrencies 61–2

strong ties 11–12
submarine warfare 134
surveillance/State surveillance 31–2, 47
    battle over privacy 38–41
    new technologies and business models 44–6
    resetting cooperation between governments and information technology companies 41–4
    rise of 'global Big Brother' 32–6
swarming (drones) 135
synergy 65–6

teachers 80, 81, 83–4, 87, 88–9
techno-crypto libertarians 55
techno-libertarians 39, 45
techno-rebels 56
teleworking and remote work 65, 67–9, 70–71
    and cities, global perspective 73–5
    digital nomads 71–3
territorial divides 93–5
terrorist threat 33
Thiel, Peter 22, 23, 24
TikTok 16, 17, 36–7, 44, 128
traditional media, time spent with per day 10
Trump, Donald 25, 119
Tucker, J.A. 126

Ukraine war 133, 135, 137, 138–9, 142–3
undersea cables 133–4
United Nations 2030 Agenda 105
urbanization 65–7

value 55, 56, 58–9, 59–60
video games 82, 97
virality 118, 127
virtual classroom 89

war 132
    artificial intelligence (AI) 136–8, 142
    change and continuity 142–3

cyberwar 139–41
digital communication networks
        133–4
drones 134–6, 137
quantum computing 141–2
weak ties 11, 12
WhatsApp 17, 46
whistleblowers 38–9

Edward Snowden 34–5, 38
Wikileaks 39
Wylie, Christopher 39

YouTube 17, 18

Zuckerberg, Mark 21–2, 24, 128–9

Titles in the **Elgar Advanced Introductions** series include:

International Political Economy
Benjamin J. Cohen

The Austrian School of Economics
Randall G. Holcombe

Cultural Economics
Ruth Towse

Law and Development
Michael J. Trebilcock and Mariana Mota Prado

International Humanitarian Law
Robert Kolb

International Trade Law
Michael J. Trebilcock

Post Keynesian Economics
J.E. King

International Intellectual Property
Susy Frankel and Daniel J. Gervais

Public Management and Administration
Christopher Pollitt

Organised Crime
Leslie Holmes

Nationalism
Liah Greenfeld

Social Policy
Daniel Béland and Rianne Mahon

Globalisation
Jonathan Michie

Entrepreneurial Finance
Hans Landström

International Conflict and Security Law
Nigel D. White

Comparative Constitutional Law
Mark Tushnet

International Human Rights Law
Dinah L. Shelton

Entrepreneurship
Robert D. Hisrich

International Tax Law
Reuven S. Avi-Yonah

Public Policy
B. Guy Peters

The Law of International Organizations
Jan Klabbers

International Environmental Law
Ellen Hey

International Sales Law
Clayton P. Gillette

Corporate Venturing
Robert D. Hisrich

Public Choice
Randall G. Holcombe

Private Law
Jan M. Smits

Consumer Behavior Analysis
Gordon Foxall

Behavioral Economics
John F. Tomer

Cost–Benefit Analysis
Robert J. Brent

Environmental Impact Assessment
Angus Morrison-Saunders

Comparative Constitutional Law,
Second Edition
Mark Tushnet

National Innovation Systems
*Cristina Chaminade, Bengt-Åke
Lundvall and Shagufta Haneef*

Ecological Economics
Matthias Ruth

Private International Law and
Procedure
Peter Hay

Freedom of Expression
Mark Tushnet

Law and Globalisation
Jaakko Husa

Regional Innovation Systems
*Bjørn T. Asheim, Arne Isaksen and
Michaela Trippl*

International Political Economy
Second Edition
Benjamin J. Cohen

International Tax Law
Second Edition
Reuven S. Avi-Yonah

Social Innovation
*Frank Moulaert and Diana
MacCallum*

The Creative City
Charles Landry

International Trade Law
*Michael J. Trebilcock and Joel
Trachtman*

European Union Law
Jacques Ziller

Planning Theory
Robert A. Beauregard

Tourism Destination Management
Chris Ryan

International Investment Law
August Reinisch

Sustainable Tourism
David Weaver

Austrian School of Economics
Second Edition
Randall G. Holcombe

U.S. Criminal Procedure
Christopher Slobogin

Platform Economics
*Robin Mansell and W. Edward
Steinmueller*

Public Finance
Vito Tanzi

Feminist Economics
Joyce P. Jacobsen

Human Dignity and Law
James R. May and Erin Daly

Space Law
Frans G. von der Dunk

National Accounting
John M. Hartwick

Legal Research Methods
Ernst Hirsch Ballin

Privacy Law
Megan Richardson

International Human Rights Law
Second Edition
Dinah L. Shelton

Law and Artificial Intelligence
Woodrow Barfield and Ugo Pagallo

Politics of International Human
Rights
David P. Forsythe

Community-based Conservation
Fikret Berkes

Global Production Networks
Neil M. Coe

Mental Health Law
Michael L. Perlin

Law and Literature
Peter Goodrich

Creative Industries
John Hartley

Global Administration Law
Sabino Cassese

Housing Studies
William A.V. Clark

Global Sports Law
Stephen F. Ross

Public Policy
B. Guy Peters

Empirical Legal Research
Herbert M. Kritzer

Cities
Peter J. Taylor

Law and Entrepreneurship
Shubha Ghosh

Mobilities
Mimi Sheller

Technology Policy
*Albert N. Link and James
Cunningham*

Urban Transport Planning
Kevin J. Krizek and David A. King

Legal Reasoning
*Larry Alexander and Emily
Sherwin*

Sustainable Competitive
Advantage in Sales
Lawrence B. Chonko

Law and Development
Second Edition
*Mariana Mota Prado and Michael
J. Trebilcock*

Law and Renewable Energy
Joel B. Eisen

Experience Economy
Jon Sundbo

Marxism and Human Geography
Kevin R. Cox

Maritime Law
Paul Todd

American Foreign Policy
Loch K. Johnson

Water Politics
Ken Conca

Business Ethics
John Hooker

Employee Engagement
Alan M. Saks and Jamie A. Gruman

Governance
Jon Pierre and B. Guy Peters

Demography
Wolfgang Lutz

Environmental Compliance and Enforcement
LeRoy C. Paddock

Migration Studies
Ronald Skeldon

Landmark Criminal Cases
George P. Fletcher

Comparative Legal Methods
Pier Giuseppe Monateri

U.S. Environmental Law
E. Donald Elliott and Daniel C. Esty

Gentrification
Chris Hamnett

Family Policy
Chiara Saraceno

Law and Psychology
Tom R. Tyler

Advertising
Patrick De Pelsmacker

New Institutional Economics
Claude Ménard and Mary M. Shirley

The Sociology of Sport
Eric Anderson and Rory Magrath

The Sociology of Peace Processes
John D. Brewer

Social Protection
James Midgley

Corporate Finance
James A. Brickley and Clifford W. Smith Jr

U.S. Federal Securities Law
Thomas Lee Hazen

Cybersecurity Law
David P. Fidler

The Sociology of Work
Amy S. Wharton

Marketing Strategy
George S. Day

Scenario Planning
Paul Schoemaker

Financial Inclusion
Robert Lensink, Calumn Hamilton and Charles Adjasi

Children's Rights
Wouter Vandenhole and Gamze Erdem Türkelli

Sustainable Careers
Jeffrey H. Greenhaus and Gerard A. Callanan

Business and Human Rights
Peter T. Muchlinski

Spatial Statistics
Daniel A. Griffith and Bin Li

The Sociology of the Self
Shanyang Zhao

Artificial Intelligence in
Healthcare
*Tom Davenport, John Glaser and
Elizabeth Gardner*

Central Banks and Monetary
Policy
*Jakob de Haan and Christiaan
Pattipeilohy*

Megaprojects
*Nathalie Drouin and Rodney
Turner*

Social Capital
Karen S. Cook

Elections and Voting
Ian McAllister

Negotiation
*Leigh Thompson and Cynthia S.
Wang*

Youth Studies
*Howard Williamson and James E.
Côté*

Private Equity
*Paul A. Gompers and Steven N.
Kaplan*

Digital Marketing
Utpal Dholakia

Water Economics and Policy
Ariel Dinar

Disaster Risk Reduction
Douglas Paton

Social Movements and Political
Protests
Karl-Dieter Opp

Radical Innovation
Joe Tidd

Pricing Strategy and Analytics
Vithala R. Rao

Bounded Rationality
Clement A. Tisdell

International Food Law
Neal D. Fortin

International Conflict and Security
Law
Second Edition
Nigel D. White

Entrepreneurial Finance
Second Edition
Hans Landström

US Civil Liberties
Susan N. Herman

Resilience
Fikret Berkes

Insurance Law
Robert H. Jerry, II

Applied Green Criminology
Rob White

Law and Religion
Frank S. Ravitch

Social Policy
Second Edition
Daniel Béland and Rianne Mahon

Substantive Criminal Law
Stephen J. Morse

Cross-Border Insolvency Law
Reinhard Bork

Behavioral Finance
*H. Kent Baker, John R. Nofsinger,
and Victor Ricciardi*

Critical Global Development
Uma Kothari and Elise Klein

Private International Law and
Procedure
Second Edition
Peter Hay

Victimology
Sandra Walklate

Party Politics
Richard S. Katz

Contract Law and Theory
Brian Bix

Environmental Impact
Assessment
Second Edition
Angus Morrison-Saunders

Tourism Economics
David W. Marcouiller

Service Innovation
*Faïz Gallouj, Faridah Djellal, and
Camal Gallouj*

U.S. Disability Law
Peter Blanck

U.S. Data Privacy Law
Ari Ezra Waldman

Urban Segregation
Sako Musterd

Behavioral Law and Economics
Cass R. Sunstein

Economic Anthropology
Peter D. Little

International Water Law
Owen McIntyre

Russian Politics
Richard Sakwa

European Union Law
Second Edition
Jacques Ziller

Regional and Urban Economics
Roberta Capello

Party Systems
Paul Webb

Evidence
Richard D. Friedman

Cultural Heritage Law
Lorenzo Casini

Federalism
*Alain-G. Gagnon and Arjun
Tremblay*